D0303066

Asbestos and Fire

Asbestos and Fire

Technological Trade-offs and the Body at Risk

RACHEL MAINES

RUTGERS UNIVERSITY PRESS

NEW BRUNSWICK, NEW JERSEY, AND LONDON

LIBRARY OF CONGRESS CATALOGING-IN-PUBLICATION DATA

Maines, Rachel, 1950–
Asbestos and fire : technological tradeoffs and the body at risk / Rachel Maines.
p. cm.
Includes bibliographical references and index.
ISBN 0-8135-3575-1 (hardcover : alk. paper)
1. Asbestos. 2. Fireproofing. I. Title.
TA455.A6M33 2005
363.738′494—dc22 2004016424

A British Cataloging-in-Publication record for this book is available from the British Library.

Manufactured in the United States of America

For Joel Tarr, who taught me to be a historian,
and who reminds us that
"new technologies create new vulnerabilities."

CONTENTS

ILLUSTRATIONS

TABLES

PREFACE

I first became interested in asbestos in 1977 while preparing a paper for the Michigan Women Studies Symposium about American needlework history between 1880 and 1930, an enterprise that has so far inspired the writing of two books, fourteen articles, and a dissertation. Among the many patterns I saw for knitted and crocheted household articles were decorative covers for the round asbestos mats that used to be available in every dime store—when there were dime stores—and used as hot mats and burner pads in American kitchens and dining rooms until the 1970s. Most of this early material was published in a lengthy paper, which, as far as I am aware, was the first scholarly article on potholders in the English language, appearing in the *Journal of Popular Culture* in the summer of 1985.

Meanwhile, I wrote a doctoral dissertation at Carnegie-Mellon University between 1979 and 1983, under the direction of urban and environmental historian Joel Tarr, on the mobilization of the American textile and apparel industries for the two World Wars, Korea, and Vietnam. Here again I encountered asbestos, reading accounts of its history as a stockpiled strategic and critical material at just the period in history at which its use had become controversial. The wave of asbestos-disease litigation, at that time soon to become a veritable tsunami, was becoming a visible public policy issue; and like any well-trained historian of technology, I noted with interest the change in public perceptions of the material.

I had other research and writing fish to fry, however, in my chosen area of interest in technology and the body, the results of which were published by Johns Hopkins University Press in 1999 as *The Technology of Orgasm: "Hysteria," the Vibrator, and Women's Sexual Satisfaction*. In 1990, I was contacted by Kathleen McDermott, then of Winthrop Group in Boston, regarding my earlier work on asbestos in potholders and as a critical material in World War II. In discussions with her and her colleagues in litigation support off and on during the next three years, I became more familiar with the medical and legal tangle that was snarling the asbestos issue and discovered to my surprise that although there were already large bodies of medical and legal literature on asbestos-related diseases and a smaller but significant body of secondary works on the history of these diseases, there was virtually no responsible scholarly history of how and why asbestos came to be used so widely. Nearly all the secondary sources I read were narrowly focused on blaming and demonizing a tiny group of individuals and businesses for a technological decision that clearly—to me, at least—had been made by the market on the advice of a large and influential group of engineering and public safety professionals. With my training in the history of technology, especially in textiles, and my interest in technology and the body, this was a topic with irresistible appeal: a textile technological trade-off issue with a health component and an almost entirely one-sided secondary literature. I continued to take notes and maintain a bibliography on asbestos through the completion of my first book in 1998.

In the past five years, my research on asbestos and fire has benefited from the advice and assistance of many persons and institutions. Joel Tarr and Kathleen McDermott stand first among those who have encouraged me to pursue this topic, making sure that I asked the kinds of questions about it that had not been asked before. Had I not met Kathleen in 1990 through Joel, it is unlikely that this book would ever have been written. Deborah Warner and Daryl Hafter, both of the Women and Technological History subgroup of the Society for the History of Technology (SHOT), have supported my efforts in textile history for more than two decades; and Deborah was kind enough to pass along to me important material from her own work at the Smithsonian Institution. This book has also benefited significantly from critiques of two versions of the manuscript by Sara Wermiel, author of *The Fireproof Building* (Johns Hopkins, 2000). Both arguments and presentation of data have been improved by her thoughtful and detailed suggestions.

I wish to thank the participants in the SHOT 2003 session on "Risky Entertainment/Pleasure Technologies," particularly the chair, Joseph Corn, and commentator David Nye. The suggestions of the audience at that meeting were useful and constructive, as were those proposed by attendees of my Hagley Research Seminar in December 2003, especially Philip Scranton, David Sicilia, Roger Horowitz, Gabriella Petrick, Francesca Guerra, and James Glynn III, among others. Although I have never met him, the lyrical and thought-provoking work of Stephen Pyne on fire has been an inspiration.

In the 1990s, much of the grunt work of assembling and organizing source material for this book fell to research assistants. Thomas Hunter spent many a weary hour in libraries chasing obscure citations and Denise Spencer many equally weary hours keying these citations and my annotations into my bibliographic data base. I was assisted in some of the early work of collecting sources by Susan Weiler, Bonnie Docherty, Grant Dixton, and Sean Perrone, all of Winthrop Group.

I learned a great deal from conversations with Scott Knowles about fire safety, and about fire meteorology from Keith Brophy Eisenman. Kraig Adler of

Cornell University was generous with information about salamanders. Catherine Gatto de Oliver played me her DVD of the Italian film *Cinema Paradiso*, with its evocative scene of a cellulose-nitrate film fire.

Like all historians, I owe a great debt to libraries and archives, the greatest of which, in my case, is to the Cornell University Library system and its dedicated staff. The patience of the interlibrary services staff and those of the Annex, Rare and Manuscript Collections, Engineering, Hotel, Law, and Catherwood (Industrial and Labor Relations) Libraries is especially appreciated. Through Cornell's participation in the BorrowDirect program, I have also made extensive use of the collections of Columbia, Penn, Brown, Princeton, and Yale university libraries. In 1996, I consulted the excellent resources of the British Library in London. The State Library of Virginia has very useful and comprehensive sources on the Richmond theater fire of 1811, and I warmly recommend the architectural library of Virginia Tech (Virginia Polytechnic Institute and State University) to anyone working in the area of theater construction. David Giordano of the National Archives made valiant efforts to locate impossibly esoteric photographs for me.

The Hagley Library and Museum in Wilmington, Delaware, made a wonderful array of materials available to me, as they do for every scholar who consults their astonishingly rich collections. The Philadelphia Free Library Periodicals Room staff pulled open many a drawer of microfilm to retrieve fifty-year-old runs of the *NPFA Quarterly*. For information on fires in their communities, I wish to thank Kathee Stahl of the Kershaw County Historical Society in South Carolina and Margaret Franklin of the Limerick County Public Library in Ireland. Deborah Warner of the Smithsonian Institution provided me with important materials from an exhibit on asbestos that she curated and guided me to the Smithsonian Institution Library's collection of more than 300,000 trade catalogs, including nearly ten linear feet of materials from asbestos-products manufacturers from the mid-nineteenth century through the third quarter of the twentieth.

I have been a member of the National Fire Protection Association (NFPA) for only two years, and I have already benefited more from this organization and its members and staff than any annual dues payment could begin to repay. All of us owe much more than we usually recognize to firefighters and fire engineers, who have given most of us the luxury of not having to think about fire very often and for saving the lives of many of us, in some cases long before we were born. In the course of this research, I browsed through more than a hundred years of publications by the NFPA and its British counterpart, the British Fire Prevention Committee, and was impressed over and over again by the striking combination of pure physical courage in the face of fire, compassionate emotional engagement with its victims, and stubborn intellectual dedication to understanding and preventing it that these professionals exhibit in their writings. On this sometimes melancholy but always inspiring journey I received valuable assistance from the able staff of the NFPA's Charles S. Morgan Technical Library, particularly Stephanie A. Naoum. I thank Dennis Berry of the Legal Department for his permission to reproduce images from the *NFPA Handbooks*.

This list would not be complete without a deep curtsey of respect to my editor at Rutgers University Press, Audra Wolfe, who has the dangerous quality of making authors feel appreciated.

As I am writing on a topic about which I have already observed that people have strong opinions, stronger in some cases than those evoked by my book about women's sexuality, and moreover one about which argumenta ad homines are at least as common as rational discussions of the substantive issues, it seems appropriate to state in advance that I have at least two biases that the alert reader is sure to note. First, I am a former business owner and employer. It has been remarked that the only experience more overrated than natural childbirth is the joy of owning your own business. The challenge of trying to meet a payroll on which the economic survival of others depends has an inevitable transforming effect. No one who has experienced this challenge can be made to believe that businesses as a class are evil or malevolent, that they seek always to maximize profit at the expense of their workers, or that business owners and managers have any real control over the market forces that push them into painful decisions. Most of all, no one who has been a business owner or closely associated with one can be made to believe that any business, regardless of size, age, creed, or country of origin, has unlimited resources and scope for altering the conditions in which it operates.

Second, I am a committed opponent of simple, unicausal explanatory theories of historical events. I find conspiracy theories particularly uncongenial because the historical record is literally paved with the corpses of failed conspiracies. For reasons that are obvious from the Prisoners' Dilemma, conspiracies are a notoriously ineffective long-term strategy. Even in the short term, as anyone knows who has ever tried to form a conspiracy to throw a surprise birthday party, somebody is sure to give the game away.

Finally, and separately from my other acknowledgments of intellectual debt, I would like to thank the family members of victims of asbestos-related disease who have listened to my story of asbestos and fire with respect, understanding that the work of their loved ones has saved the lives of thousands of others unknown to them.

Asbestos and Fire

1

The Asbestos Technology Decision Environment

The American perception of risk in the opening years of the twenty-first century has been significantly altered and reshaped by the events of September 11, 2001, and by the less deadly but still frightening episodes of Oklahoma City in 1995 and the World Trade Center in 1993. We look up at our tall buildings with a new unease; we contemplate the relative dangers of vaccination versus those of smallpox, a disease driven from the natural environment in 1980; and we regard with suspicion packages hand-addressed in block capitals, unidentified white powders, and rented trucks parked close to public buildings. Our responses to the built environment have been irrevocably altered by the framing effect of newly perceived individual and collective risks.

Between the middle of the nineteenth century and the third quarter of the twentieth, we lived with a different kind of terror: broader-based, more impersonal, and on a day-to-day level much more immediate and threatening to both individuals and communities than anything within the power of Al Qaeda or other human enemies: the danger of fire. Fire is, of course, among the oldest threats to human life; but the vast forests of North America provided what must have seemed limitless supplies of cheap building material, so abundant that even roads could be built of wood. Everyone knew how readily it burned; one needed to look no farther than one's own hearth to see the danger. Buildings burned regularly in cities, villages, and rural areas, with or without loss of life, and were rebuilt using the same combustible materials.

When populations began to be concentrated on an increasingly large scale, it became evident that something more than volunteer fire companies were needed to address the risk. First Boston and then other American cities began to regulate what could be built in urban centers, as some of their British and European counterparts had been doing since the Middle Ages.[1]

And not a minute too soon. New York, Richmond, Boston, and other American cities had all experienced major loss-of-life fires before the second half of the nineteenth century; but the epidemic of fires that swept the country after 1870 frightened citizens, business owners, builders, landlords, municipalities, and insurance companies into making common cause against a common enemy. The National Fire Protection Association was formed in 1896 and published its first handbook of fire protection the same year. The organization was part of a major international movement to establish a system of fire safety that included urban planning with the aid of building codes, telecommunication of alarms, professionalization of fire services, and fire-resistive construction with proven and tested materials, including asbestos. Nowhere was the need for such a system more obvious than in the United States, which had at the time, as it still does today, one of the highest per capita fire rates in the industrialized world.[2]

Fire in the Built Environment

On October 8, 1871, a fire driven by a massive cyclonic storm that stretched from California to Pennsylvania destroyed 17,450 buildings and the lives of 250 people in Chicago. On the same day, the town of Peshtigo, Wisconsin, north of Green Bay, lost 1,200 lives of a total population of 1,700; and in the nearby village of Williamsonville, sixty people of the seventy-seven living in the community died in an open field as wildfire consumed forest, settlement, and everything else in its path over 2,400 square miles. The Peshtigo fire killed more people than any other fire in American history, before or since. In Peshtigo and Williamsonville, which were both lumber communities, everything was made of wood—houses, streets, sidewalks, and outbuildings—and in both places sawmills added fuel; there were no structures that could provide any kind of shelter for the population. Explosions in the woodenware mill in Peshtigo hurled flaming pails, brooms, and other objects onto the heads of those who sought refuge in the river.[3] Clearly, Americans of the nineteenth century were right to fear fire.

In 1894, the lumber and sawmill communities of Hinckley, Pine City, and smaller villages in a five-county area of Minnesota were almost entirely destroyed by a forest fire that killed at least 418 persons. Trains carrying fugitives from the fire raced away over burning ties and bridge trestles, in one case with the train's wooden cars themselves on fire.[4] A few decades later, in 1918, a forest fire started by sparks from a locomotive devastated Cloquet and Moose Lake, Minnesota, and nearby areas of Wisconsin, with between 450 and 1,000 deaths.[5]

A little more than a year after Peshtigo, a major fire in Boston devastated the city's center, destroying eight hundred buildings.[6] In December 1876, a theater fire in Brooklyn killed 295 people when a hanging border caught fire, an event that was to inspire building-code clauses requiring fire-resistive stage curtains. Another theater, the "fireproof" Iroquois in Chicago, burned on December 30, 1903: within half an hour, 602 people died when the "asbestos" theater curtain protecting the audience failed to close after the scenery ignited.[7] The Iroquois was the deadliest single-building fire in American history. In 1908, the Rhoads Opera House fire in Boyertown, Pennsylvania, killed 170 persons.[8] Thanks to the diligent efforts of fire engineers and municipal authorities, these two disasters were to be the last of America's major theater fires.

Other venues were not so fortunate. On June 15, 1904, the excursion steamer *General Slocum*, filled to the gunwales with children and their parents on a church outing, burned in New York Harbor with a loss of 1,012 lives.[9] In the same year, Baltimore experienced a major fire, but, miraculously, only one person died. In San Francisco, about 3,000 people died in the earthquake and fire of 1906.[10] In March 1911, the Triangle Shirtwaist Factory fire, in the "fireproof" Asch Building in lower Manhattan, killed more than 140 workers, most of them young women, when highly combustible cotton scraps in a large waste bin caught fire in a ten-story structure with inadequate means of egress. America's worst industrial fire, it prompted significant fire- and safety-code revisions and was the last such fire in the United States to claim lives in the triple digits.[11]

Many costly and painful lessons remained to be learned, however. On Easter Monday, 1930, 320 inmates of the Ohio State Penitentiary died, locked in their cells, when a roof fire, possibly the result of arson, ignited construction scaffolding on the hundred-year-old building. Most asphyxiated in the heavy smoke. Five hundred sixteen persons were killed in Texas City, Texas, in April 1947 when two ships carrying ammonium nitrate, the explosive used by Timothy McVey in Oklahoma City, caught fire and exploded in port, setting the entire city on fire.[12]

Recreation and travel had their own risks. In September 1934, the cruise ship *Morro Castle* burned off Asbury Park, New Jersey, with the loss of 125 lives, calling attention to the combustibility of interior finishes and bulkheads on passenger ships.[13] A significant portion of the mid-twentieth-century spike in American fire deaths was attributable to entertainment and transient (hotel) occupancies: highly combustible decorations and inadequate exits were inculpated in both the Rhythm Club fire in Natchez, Mississippi, in April 1940, and the Cocoanut Grove fire in Boston in December 1942. In that era of segregation and inequality, the Natchez disaster that killed 207 African Americans did not make national headlines, but the 492 mostly white partygoers who died at Cocoanut Grove did, including many U.S. service members who were home on leave.[14]

The latter event inspired another iteration of the cycle of building-code revision and the inclusion of new requirements for fire-resistive decor and interior finishes of restaurants, bars, nightclubs, and other places newly recognized as dangerous fire environments. Two more fires in the 1940s in such venues contributed to a sense of urgency about making structures with high population densities more fire-resistive and easier to evacuate: the Ringling Brothers Barnum and Bailey Circus fire in Hartford in 1944 in which 168 died, nearly all of them children, and the Winecoff Hotel fire in Atlanta in December 1946, with 119 fatalities.[15]

Until the 1960s, American schools were a chronic threat to the lives of their students and faculty. The worst school fire in the United States occurred in March 1908 in Collinwood, Ohio, when an uninsulated heating appliance ignited wooden floorboards. One hundred seventy-four schoolchildren and two teachers perished.[16] In the aftermath, many cities upgraded their code requirements for schools and specified that heating devices and pipes near combustible materials be insulated with asbestos lagging (wrapped insulation).[17] Unfortunately, most cities, including Chicago, did not enforce these codes in existing schools.

At 2:42 A.M. on the afternoon of Monday, December 1, 1958, an alarm was called in to the Chicago Fire Department. Two minutes later, four fire engines, two trucks, and as many personnel as could be spared from other calls were sent to 909 North Avers Avenue, where children between the ages of eight and fourteen were already jumping from the upper windows of a two-story Catholic school building called Our Lady of the Angels. The fire had begun in an oil container in a basement stairwell at one corner of the building and swept up the stairways and between walls to the second floor, where children and teachers were trapped in classrooms by the dense smoke and the blazing wooden floors and wainscoting. Ninety persons, most of them children, died between 2:47 P.M. and 3 P.M. while fire fighters and police, struggling to hold back crowds of screaming and terrified parents who had flocked to the burning school, worked at desperate speed to pull children from the windows before they were overcome by smoke inhalation or caught by the flames. Five more victims died later in hospitals, bringing the death toll to ninety-five.[18]

After the fire, with funerals featuring child-sized coffins scheduled every day for weeks, fire-prevention professionals and school officials examined all the evidence from the fire and concluded that the pre-code building, constructed in 1910, was unsafe in part because its floors, walls, wainscoting, and partitions were combustible. The acoustic ceiling tiles, from which a flashover fire killed twenty-eight in one classroom alone, were made of wheat straw.[19] Municipalities again revised their codes, in some cases enforcing them in existing pre-code schools, and requiring even more extensive use of fire-resistive insulating and building materials, including ceiling tiles.[20] These fire protection efforts ultimately proved successful: it would have been cold comfort to the grieving parents of 1958, but Our Lady of the Angels was to be the last major multiple-fatality school fire in the United States.

Fire as a Weapon of Mass Destruction

At the midpoint of the twentieth century, the results of incendiary bombing in World War II had at least as great an influence on perceptions of fire risk as did disasters on our own soil.[21] The American fire protection community had watched with horrified fascination the newsreels of German bombing of British cities and the Allied response and had read reports by Sir Aylmer Firebrace, fire chief of the 343,000-member British National Fire Service in 1943, among others, on expedients such as recommending that civilians keep the water in their tubs after bathing, into which they could plunge the incendiary bombs that penetrated their roofs.[22] They had read accounts, too, by these same civilians of what it was like to live under the bombing, in constant fear of flame from the sky.[23] During the war, American fire engineers, meteorologists, insurance underwriters, architects, and other fire experts turned their professional expertise inside out by helping to plan the destruction of German and Japanese cities, drawing fire susceptibility maps that were mirror images of the maps property underwriters used to assess fire risk in American cities.[24] The attack planners were especially careful to target industrial areas with combustible roofs—the "soft targets" of incendiary bombing.

Professionals such as Horatio Bond and James McElroy of the engineering staff of the National Fire Protection Association were to find after the war, however, that nothing had prepared them for the sights, the survivors, and the statistics of the Allied air war. The effects of fire on Hamburg, Dresden, Tokyo, and other Japanese and German cities entirely overwhelmed all previous urban fire experience. Hamburg, a city of 1.7 million persons, was bombed in July 1943 by Britain's Royal Air Force. Nearly 60 percent of the built environment was entirely destroyed, most of it in a single night's firestorm, with between 60,000 and 100,000 people killed and three-quarters of a million people rendered homeless. In Dresden, where another firestorm occurred after bombing in March 1945, some estimates placed civilian mortality at 300,000, almost half the prewar population. It was impossible to determine how many had perished, since population records were consumed along with the population.[25] American visitors to Dresden, Cologne, Hamburg, and other German cities after the war were appalled by the extent of the destruction.[26]

The wooden roofs and wood-and-paper walls of Japanese residences and the incompletely fire-resistive character of even "hardened" industrial targets, combined with entirely inadequate fire-fighting capabilities, made the creation of urban conflagrations in Japanese cities a heartbreakingly easy task. In the carefully planned Tokyo firestorm of March 9–10, 1945, the fire sometimes traveled faster than a person could run. Those who took refuge in wells, basements, or other shelters were asphyxiated when the fire consumed all the oxygen as it passed over them.[27] Fatalities were estimated at between 90,000 and 130,000—

FIGURE 1.1 Scenes like this one of Hamburg after the fire storm of July 1943 impressed on American and other Allied fire-safety professionals the importance of fire-safe structures to national security after World War II.

German photo collected by the U.S. Strategic Bombing Survey.

and this was only in Tokyo. By the time the first atom bomb was dropped, U.S. bombers had destroyed almost half the built environment in sixty-nine cities in which 21 million people had lived.[28] Major Forrest J. Sanborn of the U.S. Strategic Bombing Survey said after the war, "If one needs to be reminded that civilians were the real sufferers, he need only to recall that the number of Japanese civilian casualties in the Japanese Homeland, inflicted entirely by our air force during a 6-month period, was nearly twice the Japanese military casualties inflicted by our combined Army, Navy, Air Force and Marines during a 45-month period." Sanborn did not miss the opportunity to remind his fire-professional readers that if the United States and its allies were going to dish out this kind of destruction, we would have to be prepared, in his words, to "take it."[29] The nation with the highest accidental-fire death rate in the industrialized world would have to improve the fire resistance of its communities.

Asbestos in the Fire-Safety Community

In the decision environment of highly conspicuous fire risk, a building and insulating material that was incombustible, did not spread flame, had low

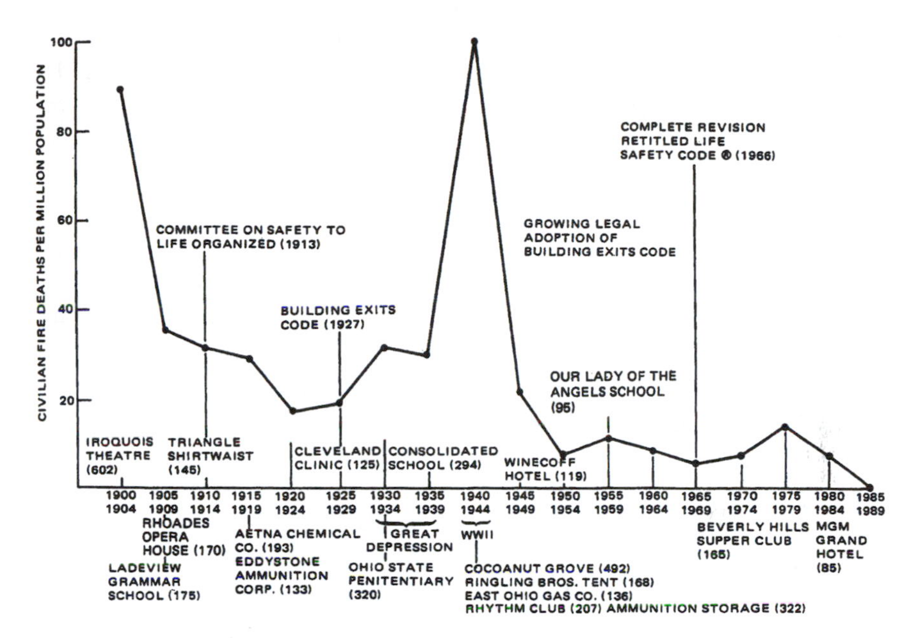

FIGURE 1.2 Except for a spike during World War II, the overall trend in twentieth-century American building fires was a steady reduction in loss of life.

Reprinted with permission from the *National Fire Protection Handbook*, 17th ed., Copyright © 1991, National Fire Protection Association, Quincy, MA.

thermal conductivity, and could be retrofitted to existing structures with well-understood technology seemed ideally suited to the goals and requirements of the evolving system of twentieth-century fire protection. Asbestos, a silica-based mineral, had been known since antiquity as a fireproof material; it is so mentioned in Pliny the Elder (A.D. 23–79), Pausanias (after A.D. 150), Plutarch (ca. A.D. 50–120), and Strabo (ca. 64 B.C.–after A.D. 21).[30] Pliny notes in *Natural Histories* XIX.iv, where he asserts that "amiantus" is a plant, that the material can be woven (with difficulty, due to the shortness of its fibers) into incombustible napkins and funeral shrouds for cremation; at XXXVII.liv, where he calls it a mineral, that it is iron-colored and mined in the mountains of Arcadia; and at XXXVI.xxxi that it is impervious to fire and useful as a talisman against witchcraft.[31] It is a commonplace in the asbestos legal literature that Pliny mentions the health hazards of asbestos, but in fact there is no such passage in any of the writings of either the elder or the younger Pliny or of any other ancient author. In his *Geography* at 10:446, Strabo notes its use in lamp wicks; Pausanias does the same in *Description of Greece* at I:26 and 27; and in *De Oraculorum Defectu* 43, Plutarch mentions an incombustible fabric made from a stone formerly (that is, before his time) mined in Euboea. Dioscorides Pedanius of

Anazarbos (first century A.D.) mentions the incombustibility of asbestos textiles. In A.D. 902, Persian writer Ibn Al Faqih noted that asbestos was sold by fraudulent talisman merchants as fragments of the True Cross. Archeological sources place asbestos mining in China in the fifth century A.D. and Iran in the eighth century A.D.[32]

Later sources, such as Swedish chemist Torbern Bergman (1735–1784), commented on the mineral's useful properties or noted the material as a curiosity from which very expensive napkins could be made, much esteemed by the ostentatious, who enjoyed amazing their friends by throwing their napery into the fire and plucking it out again, clean and unconsumed.[33] Not until the H. W. Johns Company began experimenting with asbestos in the mid-nineteenth century did the commercial possibilities of the material begin to be recognized and exploited. Johns, later Johns-Manville, first entered the market through stove and boiler insulation and roofing, but the earliest references to asbestos in building codes are to theater curtains, movie-projection-booth enclosures, and barrier board for heating appliances burning solid fuel.

Underwriters' Laboratories established in 1911 that asbestos, like pure gypsum plaster, slate, and fired ceramic, would not spread flame under any circumstances. This knowledge was later incorporated into the American Society for Testing Materials' standard for flame spread, ASTM E84, in 1958 and later established as National Fire Protection Association's NFPA 255. Asbestos-cement is still the internationally recognized zero point for fire resistance of materials and is the standard material against which all others are judged in the Steiner tunnel test of flame spread.[34] Code compliance in most communities required a flame-spread ratio of less than twenty-five; mineral wool did not and does not meet this standard; and fiberglass, while a better insulator than mineral wool, nevertheless has a nonzero ratio.

Unlike gypsum plaster and fired ceramic, which have comparable flame-spread ratios, asbestos could be woven and packed in fiber form around pipes and heating appliances and made into shingles and siding that were light enough to attach with nails to an existing building not designed for the weight of ceramic-tile roofing.[35] No other material, then or now, has these properties. The National Fire Protection Association still lists asbestos as an "excellent fireproofing agent" in its *Fire Protection Handbook*.[36]

Early in the twentieth century, insurance underwriters, safety engineers, building-code officials, and government agencies endorsed and in many cases required the use of asbestos as part of a professionally designed system of fire safety in the construction of public buildings such as theaters, hotels, and schools and in other fire-vulnerable structures such as ships.[37] In the case of buildings, this interest in fire-resistive construction continued a trend begun in the nineteenth century, when the use of brick, plaster, metal, and stone were

TABLE 1.1

Flame-Spread Ratios of Fire-Resistive Materials

Material	*Flame-Spread Ratio*
Asbestos	0
Ceramic/brick	0
Fiberglass	10–20
Mineral wool	30–40

thought to be adequate safeguards against building fires. The experience of Boston's 1871 fire, in which granite exposed to high temperatures deteriorated and "melted," and later fires, in which sprinklers proved unsatisfactory, steel supports weakened and collapsed, brick softened, and concrete cracked and spalled, reinforced the growing conviction that buildings and their occupants could not be reliably protected from fire without asbestos.[38] After 1900, because of municipal building codes, it was increasingly difficult to build any structure inside the fire limits of any American municipality—the highly built-up or downtown area—without using at least some asbestos.

Outside those limits, because of insurance underwriters' codes like that of Factory Mutual, it was difficult to insure and therefore to finance any structure that did not have asbestos in the applications for which it was recommended or required by the code.[39] Banks would not (and will not) mortgage an uninsured building. Of the eight types of built-up roof coverings approved by the Underwriters' Laboratories (UL) in 1935, for example, seven included asbestos.[40] The UL standards, as set forth in various model building codes, including those of the National Board of Fire Underwriters, were in many cities simply adopted wholesale into municipal building codes. Some communities even held property owners liable if burning brands from their wooden roof coverings ignited nearby structures, a strong incentive to use fire-resistive roof coverings.[41] A large area of Salem, Massachusetts, had been destroyed in 1914 after a single wooden-shingle roof fire spread burning brands for blocks in every direction, creating a scene of devastation used for decades in advertising booklets by the National Board of Fire Underwriters (see fig. 1.3).

Fire was an immediate and visible risk; asbestos was not. Partly due to limitations of epidemiological knowledge and testing instrumentation, the long-term effects of low doses of asbestos were not perceived until several decades

FIGURE 1.3 The Salem, Massachusetts, conflagration of June 1914 was regarded by fire underwriters in the United States and Britain as a classic cautionary example of the dangers of combustible roofing materials. This image, which was widely used in fire-insurance publications, was the first disaster photograph ever taken from an aircraft.

George Murray, for the *Boston Journal*, 1914. Original in the National Archives.

after the beginning of widespread asbestos use.[42] The fire safety community was not, in any case, focused on long-term risk or on hazardous materials other than those that either contributed to fires and explosions, such as gasoline and grain dusts, and those which emitted toxic fumes when heated, such as nitrocellulose film. Their jobs were and are to prevent fire and to prevent loss of life and property when fires occur.

Bodies at Risk: The Asbestos Controversy

In the decades since the passage of the Occupational Safety and Health Act (OSHA) of 1970, asbestos has become a highly visible and controversial legal, scientific, and public health issue, with hundreds of thousands of exposure cases creating a virtual subindustry of torts. Thousands of hours of court time and millions of work hours, not to mention billions of dollars in settlements and judgments, are expended annually in resolving issues related to asbestos. The Rand Institute for Civil Justice reports that asbestos litigation has so far cost businesses about $40 billion, and another $210 billion in costs are possible.[43] The $50 billion costs of asbestos removal—and torts resulting from it—also add to the total, as does the $1 trillion in depreciation of asbestos-containing buildings. The burden on the courts is astonishing: by 1989, asbestos cases were 3.52 percent of all civil cases, and 61 percent of all product liability cases in the federal courts.[44] Because the United States is the only industrial democracy in the world without national health insurance, the minority of asbestos plaintiffs who are actually sick have no recourse but litigation to defray the cost of care, thus making us also the only industrial nation that has permitted asbestos disease to become an apparently uncontrollable drain on our own economy as well as that of other nations.[45]

Despite the public attention the controversy has received, there has been surprisingly little discussion of how and why asbestos was introduced into the built environment in the first place or of the risk perceptions and cost-benefit analyses that resulted in Americans' use of an average of five to ten pounds of asbestos per person per year by the 1960s.[46] None of the major works on asbestos risk cites any of the professional literature of fire protection; indeed, reading these works, one would not guess that such a literature exists or suppose that the opinion of any community of experts other than health professionals might be relevant to the debate. In an article included in Barry I. Castleman and Stephen L. Berger's *Asbestos: Medical and Legal Aspects*, Berger, for example, argues that alternatives to asbestos insulation have existed since 1890, citing as evidence a list of patents for products of which fewer than 10 percent were ever manufactured or tested. He declines even to mention the obvious necessity of testing materials for their code compliance or to suggest how any of the supposed substitutes might have fared on these tests.[47]

Like nearly all new technologies, asbestos was evaluated between 1870 and 1964–65 against a background of the problems it solved, not of those we were later to learn that it created. The negotiation of standards of safety in the United States has historically involved complex interactions among a variety of constituencies, including consumers, government, industry, and the community of experts, which typically include scientists, engineers, and physicians. The latter groups are by nature conservative, assimilating new knowledge slowly over a period of years or even generations. The contagion debate of the nineteenth century was just such a controversy.

Perceived risks have been the impetus for the development of many kinds of technology, and it is not unusual for technologies to be developed that successfully reduce the originally perceived risk but that are ultimately found to be flawed with risks of their own. Infant seats in automobiles are a recent case in point: they save adult lives but endanger small children. In the case of asbestos, the perceived risk was fire, an obvious and unmistakable public health hazard. The United States, like all industrial democracies, attempts to base most policies on the principle first articulated by Jeremy Bentham, that of the greatest good for the greatest number. In 1948, of the approximately 10,000 Americans who died from fire every year, almost 40 percent were children—about ten deaths per day of children of elementary school age or younger.[48] In 1999, the latest year for which the Centers for Disease Control (CDC) have published statistics, 2,355 persons died of mesothelioma and 449 of asbestosis, none of whom were under the age of fifteen. More *children* died every year from fire, before we built the fire safety system that includes asbestos, than *adults* are now dying from asbestos-related disease.[49]

Asbestosis is a respiratory disease caused by inhaling asbestos fibers, resulting in scarring of lung tissue, which gradually loses its elasticity. Length and intensity of exposure usually determine the severity of the disease. This is not always the case with mesothelioma, however, a type of cancer that produces tumors on the pleura (lining of the lungs) or the peritoneum (lining of the abdomen); neither length of exposure nor the intensity—the concentration of fibers in the inhaled air—are well correlated with the incidence of the disease, which has even been known to occur in individuals with no history of asbestos exposure. Miners in minerals found in association with asbestos, such as talc, can develop both diseases.[50] Latency periods are long, up to thirty years, a characteristic of asbestos disease that has greatly complicated traditional systems of workers' compensation and other insurance. These programs were designed to cope with immediately obvious injuries and toxic effects, not those that do not manifest for two or three decades. In some states, workers must file claims within a period shorter than the latency period of mesothelioma, resulting in a flood of claims from the so-far unimpaired. The number of claims exceeds the annual number of deaths from both diseases by two orders of magnitude.

Asbestos building materials and the system of fire safety of which they were an important component successfully reduced the annual rate of fire deaths in the United States from 9.1 per 100,000 in 1913 to 1.0 in 1998, during which period asbestos became one of the dominant technologies in structural fire prevention.[51] Only after decades of widespread asbestos use did a new generation of epidemiologists begin to shift the attention of the public and of policymakers from the problems solved by the technology to those it apparently caused. According to William W. Lowrance in *Of Acceptable Risk*, we could not have known the effects of asbestos before a period of long exposure to it.[52]

The framing, or context, of asbestos risk perception was altered by the apparent removal of the background of fire danger. A. Tversky and D. Kahneman have shown that the framing of risk-benefit analysis can completely reverse preferences, even when the mathematically calculated utility of each outcome is identical.[53] Asbestos was evaluated between 1900 and 1964 against a backdrop of scenes such as that of the military and civilian bodies found in the ruins of the Cocoanut Grove nightclub in 1942 and the rows of children's bodies in the Chicago morgue after the Lady of the Angels school fire in 1958. Due in part to the widespread use of asbestos in construction between 1930 and 1980, large loss-of-life fires are now very rare, and the technology of asbestos is evaluated today against the backdrop of cases of mesothelioma and asbestosis in adults.[54]

Our answer to the policy question "Do we want to use asbestos?" has changed because a new scientific paradigm for asbestos has emerged since 1965, within a frame in which fire risk has become a relatively minor concern compared with 1900. We did not have the information in 1900 or even in 1950 to weigh the value of young lives potentially lost to fire against possible cases of cancer and mesothelioma in adults a half century later, and it is possible that our decision would have been no different if we had. Given the possibility of such a choice, we would obviously prefer to have neither fire deaths nor cancer, but history seldom permits the luxury of straightforward comparisons. Decisions to use asbestos and decisions to restrict it were separated by many decades. It is highly improbable that any alternative fire-prevention technology we might have developed would have been risk-free; in fact, fiberglass and mineral wool, both proposed alternatives to asbestos, are suspected health hazards.[55]

Two different kinds of risk are at stake here, which are separated temporally, conceptually, and usually spatially as well. The first is the acute, obvious, and short-term risk of fire, which kills in minutes or even seconds and may end the lives of many persons at once, even whole communities of thousands of individuals. No scientific expertise is required to judge the danger of fire; even animals understand it, and its potentially fatal effects are nearly always immediate and perceptible, not long-term and latent. In the face of an actual fire, it is not possible to lose what Ulrich Beck calls one's "cognitive sovereignty" over the perceived risk of fire.[56] Moreover, individual susceptibilities are of little significance once ignition occurs, although the individual may greatly increase the likelihood of ignition of his or her immediate environment by engaging in risky behaviors such as installing a solid-fuel stove on a wooden floor or smoking in bed. The increased risk to smokers is in fact one of only two elements that fire and asbestos-related disease have in common.

The second is one that asbestos-related disease and fire share with nearly all risks: until it is obvious that the worst has occurred, potential victims are inclined to think that disasters of the kind in question only happen to someone else. Thus, the owners of the Station nightclub in West Warwick, Rhode Island, assumed in February 2003 that because fireworks had been used on their stage many times before without incident, no incident would ever occur; and asbestos workers with decades of experience discounted or ignored advice to wear dust masks. All of us perform these acts of denial every time we drive our automobiles, an activity that poses far and away the greatest risk to life and limb that most of us experience on a day-to-day basis. Everyone knows that obesity greatly increases the risk of disease and death, but more than a third of Americans are overweight. Most of them will lose weight only when it is apparent that fat is killing them, if then.

The risk of asbestos-related disease is, of course, very different in almost every other respect from that of fire. We do not reflexively fear dust floating in the air, even if it is visible, which it often is not, as we do the proximity of fire. The effects of inhaling asbestos fibers are long-term and chronic, not short-term and acute, and they occur to individuals differently and separately. While groups of individuals may be breathing the same air and inhaling the same dusts at the same time, only a portion of the group will contract either asbestosis or mesothelioma or both, and members of these subgroups will not all show the same symptoms. If active disease occurs after exposure, which it does in only a minority of those exposed, it does so after unpredictable lapses of time: sooner in some individuals, later in others. All of us in the United States have been exposed to asbestos fibers, and those of us who live in urban areas (even small towns) inhale at least a few of them every day of our lives, but we do not all have asbestosis or mesothelioma. Thus, the conceptual link between what one is doing at the moment and the risk of dying from it seems much more remote than in the case of the stove owner whose wooden floor catches fire.

Moreover, because there is no known threshold limit value (TLV)—that is, no known safe level of exposure to asbestos inhalation—science can offer no plausible probabilities for an individual who is environmentally exposed to asbestos. Even for those who are occupationally exposed, disease more often follows years of high-level exposure than hours or weeks of low-level exposure; but there are so many counterexamples that some scientists, unable to determine a safe limit, assert that the safe limit is an impossible zero exposure. In

1971, the National Academy of Sciences was still expecting to discover a safe lower limit, but to date none has been convincingly demonstrated.[57]

Finally, it is impossible to know what dangers one has escaped. Those who die in middle life or later of asbestosis or mesothelioma have no way of knowing whether or not they are among those who might have died in fires had not the buildings in which they lived, worked, and were educated had systems of fire safety that included asbestos. At the individual level, the two kinds of risk are incommensurable.

At the level of policy, as well, the two risks are difficult to compare in the United States, but other nations seem to be having more success in this regard. In most industrial democracies, asbestos-related disease is regarded as one of many possible dangers to which citizens may be occupationally or environmentally exposed; and treatment and support are available regardless of how, when, or by what agency the individual was exposed. The costs of injury from asbestos, like those of fire, traffic accidents, lead poisoning, and all other human afflictions, are distributed over the whole taxpaying population. It is clear that the entire population benefits from fire prevention, and asbestos-related disease is regarded as one of the expected costs of developing fire-prevention technologies, even as asbestos-based methods of making structures fire resistive are phased out. In a national system with universal health care, however administered, both occupational and environmental diseases are treated as "normal risks" in the sense that Charles Perrow writes of "normal accidents."[58]

In the United States, we accept collective responsibility for the political, technological, and market forces that adversely affect the health and well-being of individuals only in very specialized cases, such as that of the military and, after much political agitation, of coal miners. In the case of manufactured products, we have often substituted litigation for social support, addressing the responsibilities of technological choice, in Niklas Luhmann's words, by trying to "discover scapegoats, and to consecrate the victims of unheralded calamities."[59]

Asbestos-related disease has been treated in law as a "deviant risk," something imposed on the unsuspecting (and apparently incurious) by a cabal who allegedly arrogated the results of science to itself rather than as the result of ordinary technological change in an industrial democracy.[60] This policy has become a disaster unto itself, consuming billions of dollars, not in health care but in litigation and lost jobs. We have, in the case of asbestos, abdicated our responsibility for support of those injured in the necessarily experimental process of learning to protect ourselves from fire to the lawyers and the courts. As noted, asbestos litigation and asbestos bankruptcy are lucrative subindustries within the colossal mega-enterprise of the American legal profession. Isabel Allende points out that three-quarters of all the lawyers on earth practice in the United States.[61] Among the lot of them, in the absence of any responsible national health policy, they have created a legal conflagration of epic proportions around the risk and health issues of asbestos technology.

Most components of our fire-safety system, such as sprinklers, are uncontroversial. As a species that lives and works voluntarily with fire, we have of necessity developed a variety of technologies for containing it. Around 1900, the old danger from fire was newly recognized from relatively recent sources: electricity and combustion engines and their fuels, especially automobiles, ships, and airplanes.[62] Because asbestos protected effectively from the highly visible and acute risk of fire, the use of asbestos expanded greatly during the first seven decades of this century.[63] Asbestos use in the United States rose from less than half a million pounds in 1937–39 to about 1.2 billion pounds per year in the 1940s and reached 1.5 billion pounds in 1950, more than five pounds a year for every man, woman, and child then living in the United States.[64] Most of this asbestos is still in place, protecting us from fire but exposing us, we are told, to long-term cancer risks.

Response to fire risk was the motivation behind the U.S. Navy and British Construction Rules' specification that ships be built with asbestos insulation. Ship fires, both accidental and as a result of enemy action, were a significant cause of death and serious injury not only on navy ships but on merchant and passenger vessels.[65] In addition, the use of asbestos lagging and insulation in ships and other defense applications involving steam pipes during World War II and the Korean War greatly improved the efficiency of a strategic commodity, fuel.[66] Asbestos insulation provided a fire-safe means to reduce fuel consumption by more efficient energy production, thus not only conserving wartime energy stocks but also reducing pollution and dependence on imported petroleum. Of the conservation of fuel, the *Asbestos Worker* had written in 1927 that insulation on one hundred square feet of pipe saved 150 pounds of coal every ten hours. If pipes were not insulated, "At the rate of three hundred working days per year, this means an annual waste of forty-five thousand pounds of coal, or twenty-two and a half tons for each one-hundred square feet of unprotected pipe," not including boilers and valves. A shipbuilder wrote in 1936 that "[o]ne linear foot, uninsulated, of an 8-inch line carrying steam at 250 pounds and 60 degrees superheat, will waste at least 15,000,000 British thermal units in a 5,000-hour operating year; and a standard weight 8-inch flange, if bare, 2.4 times as much."[67]

The considerations that motivated defense planners to add asbestos to the textile materials on the Strategic and Critical Materials (SCM) list in 1940 and to keep it there until stockpile objectives were met in the early 1960s were fuel conservation, the dangers posed by fire to U.S. service members, and the national security risk of running out of asbestos for vital defense needs such as insulation, brake linings, clutch facings, gas mask filters, and other defense applications. During World War II, and to a lesser extent in the Korean conflict, the government purchased large supplies of asbestos for the SCM stockpile and released them to manufacturers working on defense orders according to a system of priorities and allocations. During this period, not even the asbestos workers' union or the U.S. Department of Labor called attention to any unusual risks associated with the manufacturing or handling of asbestos products.[68]

Defense concerns continued to dominate federal policy for asbestos until the 1980s, at which time the government was still selling off its defense stockpile of asbestos.[69] The Bureau of Mines was especially reluctant to accept the new paradigm of asbestos risk, since it had been for more than half a century a strong supporter of the U.S. asbestos industry.[70] The U.S. Navy also clung to the advantages of asbestos as a ship fireproofing material and remained the largest American user of asbestos as late as 1969.[71]

The effect of eliminating asbestos from ship construction has not, in fact, been entirely salutary. W. Keith C. Morgan and J. Bernard L. Gee assert that

> the number of casualties caused by burns in the Royal Navy warships during the Falklands War, and on the USS Stark when the latter was hit by an Exocet missile in the Persian Gulf, was appreciably greater than expected because of the exclusion of asbestos insulation from these ships. One third of all Royal Navy casualties in the Falklands War had burns, and in the overall campaign, burns accounted for 14% of all injuries. The comparable figure in the 1939 to 1945 war was 1.5%. Thus, burn casualties were almost 10 times as common in the recent conflict in which relatively inflammable materials, such as aluminum, were used in the construction of the ships, and in which asbestos had not been used for fire control. Nor should it be forgotten that the Challenger disaster was a consequence of substituting a nonasbestos-containing putty used to seal the O rings of the craft.[72]

Fire-safety considerations prompted school and other construction planners to include asbestos building materials in the wave of building that accompanied the birth of the baby boomers after World War II. Parents, teachers, and administrators were justifiably appalled by the school fire rate in the early 1950s: seven to ten per day, on average, in 1951.[73] In a fire-prone decision environment, what few risks were then perceived from asbestos seemed negligible compared with the appalling losses of life and property from fire.

The fifties were characterized by a chronic shortage of asbestos building materials to meet these new demands for fire-resistive construction.[74] When asbestos was discussed in a safety context in the professional management literature of the 1950s, it was nearly always in the context of preventing fire and fire casualties.[75] Asbestos employment went from 3,000 in 1930 to 26,000 by 1970.[76] The short-term results were immediately visible and encouraged further use of the material: per capita fire deaths in America peaked during World War I and then began falling steeply through the twentieth century, except for a spike during World War II.[77] The rate continued to decline, both in absolute terms and in rates per 100,000 people, until 2000, after twenty years of asbestos "abatement," when it rose an ominous 20 percent in two years, from 1.0 per 100,000 in 1998 to 1.2.[78] Between 1999 and 2001, a few more than 3,300 Americans per year have perished by "accidental exposure to smoke, fire, and flames." Obviously, figures for fire deaths in 2001 compiled by organizations such as the National Fire Protection Association are much higher because they include the victims of the 9/11 terrorist attacks; but since these were not accidental, they are not so classified by the CDC.

Fire death statistics are compiled in a number of ways for different purposes. At this writing, the most complete register of fire deaths in the United States is that of the National Fire Protection Association, which compiles its data from state and local governmental records. The CDC figures are much more conservative than those of the NFPA because they include immediate but not proximate causes. Death by leaping from the window of a burning hotel, for example, would be classified by CDC as a fall, whereas NFPA would list such a death as having been caused by fire. There are other differences as well: to the NFPA, a "civilian" is someone who is not a firefighter; to the CDC, a "civilian" is someone who is not in the armed forces. The 2001 deaths in the terrorist attack were variously classified by the CDC according to the ways in which the states involved issued death certificates, but most were listed as homicides; NFPA included all the fatalities as having been caused by fire and explosion.[79]

Hundreds of thousands of people are now alive at least in part because the buildings or other structures in which they lived, went to school, or worked were rendered fire resistive by asbestos. The difference between the current fire death rate and that of 1942 amounts to a fifty-three-year average of about 7,500 lives annually, or about 400,000 lives. The 1942 rate actually represents an improvement; as the 1991 edition of the NFPA *Handbook of Fire Protection* notes: "Fire deaths have fallen by roughly 50 percent in the 70 years since their peak levels around World War I. Fire death *rates* have fallen by three-fourths."[80]

Not until the mid-1960s did asbestos enter the last step of risk assessment: the discovery of new danger in a technology developed to cope with an older danger. As the per capita rate of fire mortality fell, the visibility of fire as a risk factor was greatly reduced. By the end of the sixties, asbestos exposure was the more visible of the two risks.[81] While history cannot record what did *not* occur (fire and explosion deaths on American ships and in schools, homes, theaters, nightclubs, and restaurants prevented by the system of fire safety that included asbestos materials), it does record injuries caused by asbestos exposure.[82] It is only in hindsight, however, that we have been able to assess this long-term risk and only in the last three decades that we have resumed the iterative, and almost certainly futile, search for a risk-free fire-protection technology.

TABLE 1.2

U.S. Fire Deaths, Asbestos Use, and Population, 1932–80

Year	*U.S. population*	*Fire deaths (conflagration and burns)*	*Fire deaths per 100,000 people*	*Asbestos use (in pounds)*
1932	124,949,000	6,913	5.6	197,212,000
1937	129,969,000	7,928	6.1	504,000,000
1941	133,402,000	6,963	5.2	877,482,000
1942	134,860,000	7,920	5.9	868,242,000
1943	136,739,000	10,450	7.6	892,000,000
1944	138,397,000	10,040	7.2	778,482,000
1945	139,928,000	9,170	6.5	756,060,000
1946	141,389,000	8,790	6.2	917,454,000
1947	144,126,000	8,940	6.2	1,233,586,000
1948	146,631,000	7,688	5.2	1,356,888,000
1949	149,188,000	5,982	4	1,070,264,000
1950	151,325,798	6,405	4.2	1,457,606,000
1951	154,360,000	6,788	4.4	1,458,000,000
1952	157,028,000	6,922	4.6	1,505,218,000
1954	162,417,000	6,083	3.8	1,054,244,000
1957	171,229,000	6,269	3.7	1,446,984,000
1958	174,141,000	7,291	4.2	1,112,748,000
1959	177,073,000	6,898	3.9	1,516,090,000
1961	179,323,175	7,102	3.9	1,331,088,000
1965	193,526,000	7,347	3.8	1,589,533,100
1970	203,051,000	6,718	3.3	1,468,278,900
1975	213,051,000	6,071	2.9	197,308,000
1980	226,542,203	5,822	2.6	791,459,650
1985	238,466,000	6,313	2.6	357,148,865
1990	248,709,873	5,302	2	91,156,736

Sources: U.S. Bureau of the Census, NFPA, and U.S. Bureau of Mines.

The Health Risk Debate

The regulatory climate of the 1950s and 1960s reflected the overall context of risk perception. State laws and local ordinances mandated inspections for obvious hazards such as fire, explosions, unprotected machinery, and overloading of floors or other supports. Inspectors in most states and municipalities looked for "acute" risks such as anthrax in woolen mills and brucellosis in dairy plants.[83] As Barry I. Castleman and Grace E. Ziem observe, "There was no federal regulation of general industry workplace hazards until 1971; and state and local agencies were thinly staffed and minimally funded."[84] The concept of long-term or chronic risk was very poorly understood except in connection with heavy metals such as arsenic, mercury, and lead and with mineral extraction and other "dusty trades."[85] Limitations of the instruments available for testing also impeded the progress of knowledge of industrial hygiene. When Pittsburgh elected to clean up its fatally polluted air at the end of the 1940s, for example, the only instrument available for testing air quality was the Ringelmann chart, usually represented by a strip of celluloid marked with dark shadings from gray to black, to be held up in the general direction of the sky for comparison: dark, darker, and too dark.[86] Similarly primitive instrumentation prevailed in other areas of environmental testing. Dust hazards such as asbestos could only be identified if the dust were actually visible by existing means, and air content could not be readily analyzed until electron microscopes became commercially available in 1965.

David Ozonoff seems to be arguing in his 1988 article "Failed Warnings" that we should have known better than to use asbestos.[87] The principal difficulty with Ozonoff's hypothesis is that it is ahistorical in the contexts of both regulatory history and the history of science and medicine. As we have seen, regulation rarely occurs in advance of public recognition of risk and public acceptance of the restrictions of private liberties that such regulation requires.[88] In the case of lead, for example, the scientific community recognized risk in the first decades of this century; but lead was not banned from paint and gasoline until the 1970s, and lead arsenate continued to be used as a pesticide on fruit for decades after scientists first suggested that the substance posed a threat to the health of consumers.[89] Usually, although not invariably, it takes a fatal or near-fatal disaster on the scale of a cholera epidemic, the Iroquois Theatre fire, the Gauley Bridge Tunnel, Love Canal, Three Mile Island, or Chernobyl to focus public attention on the danger. In the case of asbestos, it was the publicity associated with the October 1964 meeting of the New York Academy of Sciences, which for the first time brought the hazards of asbestos to a public larger than the community of occupational health professionals.

In the case of asbestos, the medical literature has reflected the traditional conservatism and rhetorical restraint appropriate to the publication of scientific findings. In addition, like nearly all scientific literatures of long-term discovery, it contains misinterpretations of results, experimental false negatives (and probably false positives as well), observer effects, mistaken causes, and results that were misleading due to limitations of instrumentation.[90] Appropriately, within the protocols of the scientific method, it contained a number of calls to additional research, particularly to expensive longitudinal studies, to replicate and confirm observations of samples as small as one member.[91] Occupational-disease researchers such as W. C. Hueper, who called for immediate action on asbestos before 1964–65, were subjected to professional ridicule as unscientific alarmists, going off half-cocked before all the scientific data were sorted out and confirmed.[92] The publication of the New York Academy of Sciences conference papers in 1965 marked the end of this exploratory phase of scientific perceptions of asbestos risk.[93]

The literature of the period between the first British observation of pathological symptoms in asbestos workers in 1898 and the 1965 New York Academy of Sciences report can be loosely categorized into four types: definite warnings of risk within the constraints of legitimate scientific skepticism, inconclusive studies that include calls for research into possible hazards, findings of no risk, and what I have called red herrings—scientific reports that mistook the cause of the observed symptoms or suggested solutions to the problem that ultimately proved ineffective. All of these types, as I have indicated, are characteristic of how science explores a new problem space. There is, of course, considerable overlap among them; for instance, some of the definite warnings include red herrings. What follows are examples from this literature rather than an exhaustive survey.

Factory inspectors in England noticed in 1898 and 1906 that workers in dusty trades, including asbestos workers, experienced respiratory problems related to their work.[94] The first recognized death from asbestos exposure, a patient under the care of Montague Murray at Charing Cross Hospital, occurred in 1900.[95] In July 1924, W. E. Cooke published a very brief research note in the *British Medical Journal* about a woman who had died after working in one of the dustiest area of an asbestos textile mill, the card room, for thirteen years. In the 1927 follow-up to this article, he named her disease asbestosis and provided further details on her demise. The dust in the textile mill had been so thick that workers in the card room "could not see each other." Cooke, who can be regarded as the originator of asbestosis as a disease paradigm, thought that improved ventilation would have prevented her illness.[96] In 1930, asbestosis became a compensable illness in Great Britain, and in the United States, references to articles on asbestosis began to appear regularly by name in the *Industrial Arts Index*.[97] W. B. Wood and S. R. Gloyne expressed an opinion similar to Cooke's in a *Lancet* article in 1934 and suggested that there would be no new cases of asbestosis in properly ventilated factories.

In the United States and Britain in the 1930s, studies of asbestos workers tended to focus on two red herrings: tuberculosis and silicosis. Asbestosis was thought by many to be a predisposing cause of tuberculosis. Writing in the *Journal of Industrial Hygiene* in 1933, Philip Ellman comments that

> Merewether, in 1930, found only 4 active cases of tuberculosis in an examination of 374 asbestos workers actually at work, and formed the impression that there was not outstanding susceptibility to tuberculosis among these workers. On the other hand, Wood and Gloyne in 1931 were able to trace 12 cases of tuberculosis, 10 of which were active, in a series of 57 cases of asbestos; and among 35 deaths from asbestosis referred to in the [British] Report of the Chief Inspector of Factories for 1931, tuberculosis was a complicating or terminal factor in 11 cases. Among seventeen of my own definite cases of asbestosis, the majority of whom were not at work, six have tuberculosis, of which four were active.[98]

Note that Ellman, in responsible scientific style, calls attention to results that differ from his own (Merewether's) and emphasizes the tentative and exploratory character of research results in his subject area. When Ellman wrote, the attention of occupational medicine had been for three decades focused on tuberculosis as the most common and most often deadly disease of industrial workers.[99]

Philip E. Enterline ably documented the process of changing scientific opinion in a 1991 article in the *American Journal of Industrial Medicine*. He summarized his results as follows:

> Literature published in the years 1934–1965 was reviewed to determine attitudes and opinions of scientists as to whether asbestos is a cause of cancer. In Germany, the issue was decided in 1943 when the government decreed that lung cancer, when associated with asbestosis (of any degree), was an occupational disease. In the United States, however, there was no consensus on the issue until 1964. Opinions of scientists over a 22 year period are shown and the contributions of various cultural, social, economic and political factors to these opinions are discussed. A lack of experimental and epidemiological evidence played a major role in delaying a consensus. Other important factors included a rejection of science conducted outside of the U.S. during this period, particularly a rejection of German scientific thought during and after WWII, and a rejection of clinical evidence in favor of epidemiological investigations. Individual writers rarely changed their minds on the subject of asbestos as a cause of cancer.[100]

While U.S. medical experts of the 1940s were reluctant to rely on German sources on the health dangers of asbestos, another significant class of trained

American professionals, as we have seen, was learning important lessons from Germany and Japan. These were the engineers, insurance underwriters, architects, firefighters, and meteorologists whose risk perceptions were indelibly imprinted with the nightmare images of fire as a weapon of mass destruction. Of this experience, the editor of the *New Yorker* wrote in December 1943 after the bombing of Berlin:

> Once again, in their collective reaction to the destruction of Berlin, New Yorkers have demonstrated that they don't know what fear is, any more than they know what a number of other fundamental passions are. In the past week or so, many people have expressed opinions with which we have agreed. We agree with our barber that it serves the bastards right; we agree with Major Eliot, that it was a necessary action, efficiently and economically carried out, of vast military and political significance; we agree with Walter Lippmann that each individual should shoulder his share of the moral responsibility for it, and do what he can to make sure that his children's consciences will not have to be burdened as his is. One implication, however, is still, as far as we know, fluttering around untrapped. Nobody has pointed out that the destruction of Berlin established the fact that it is now possible to destroy a city and that every city, but for the hairline distinction between the potential and the actual, is afire, its landmarks gone and its population homeless. From where we sit, the flames are clearly visible.[101]

From where the fire safety professionals sat, the flames were very clear indeed.

2

Asbestos before 1880

From Natural Wonder to Industrial Material

Asbestos is an old material, first used by humans in the Neolithic Age as a temper for ceramics. Prehistoric shards and ware containing asbestos have been found in Finland, central Russia, and Norway, and at Lapp sites in Sweden; and the material was still being used for this purpose in the 1970s in Uganda and Kenya. Native American sites have yielded asbestos fragments among other Indian artifacts such as bone tubes and scraping tools shaped from antler.[1] It has been known since antiquity in both the western world and Asia as a natural wonder and a source of fiber for very expensive, and therefore rare, textile objects such as shrouds, napkins, tablecloths, and special-purpose clothing. In the west it is first mentioned in Greek sources—thus its name, ἄσβεστος, "unquenchable." Where the word appears in the Homeric corpus and the New Testament, it is always used as an adjective referring to this quality, not to the mineral, as in unquenchable laughter or fame.[2]

Between antiquity and the middle of the nineteenth century, asbestos made a slow and irregular transition from a rare and miraculous substance: a fiber from stone from which a fabric could be produced that would not burn, through various misunderstandings of its nature and properties, to a systematic integration of the material into modern chemistry and mineralogy. Because its fire and acid resistance seemed almost literally too good to be true, descriptions of it in the Middle Ages and even well into the early modern period have a mythical quality that would do justice to Ripley's "Believe It or Not," as we shall see. By the end of the eighteenth century, however, scientists had discovered the mineral's usefulness in filters and for fire resistance, and experiments had begun that put asbestos on the road to becoming an industrial material by 1850.

The first known western reference to the mineral asbestos may be Theophrastus' *De Lapidibus* [*On Stones*], written about 300 B.C. by Aristotle's student and successor at the Lyceum in Athens. Marcus Terentius Varro

(116–29 B.C.), in his *De Lingua Latina* [*On the Latin Language*] lists fabrics that are *asbestinon* (fireproof) as among those that *post luxuria attulit*, "which extravagance brought at later times," and notes that the word is from the Greek.[3]

The only naturally occurring mineral that can be spun and woven, asbestos is actually not a single mineral but a group of asbestiform (fibrous) metamorphic rocks usually found in veins with other minerals. A chain silicate, its fibers can be separated along the crystallographic planes of its molecular structure. Chrysotile, or white asbestos, is a serpentine (hydrous silicate of magnesium) found historically in India, Siberia, and the Mediterranean area and in modern times in Canada, Russia, southern Africa, China, and parts of the United States.

The amphibole group consists of tremolite, amosite, crocidolite (blue asbestos), actinolite, cummingtonite, grunerite, riebeckite, and anthophyllite. Tremolite and actinolite are chemically the same mineral (in pure form, $Ca_2[Mg,FE]_5Si_8O_{22}[OH]_2$). Chrysotile ($[Mg,Fe]_3Si_2O_5[OH]_4$) has longer and more flexible fibers than do amosite types, resists abrasion, and is the material from which most asbestos fabrics are woven.[4] All types of asbestos are chemically inert, have low thermal conductivity, and resist fire and acids, the last three of which qualities were noted by ancient Greek and Roman writers.[5] Some anthophyllites, tremolites, and amosites are unaffected by temperatures up to 5,000 degrees Fahrenheit; chrysotile is embrittled by exposure to very high heat; and crocidolite, as mineralogist Oliver Bowles points out, "is easily fused into a black magnetic mass" due to its high iron content.[6]

Asbestos in Antiquity

In chapter II.17 of Theophrastus' *On Stones*, the Athenian scientist (372/1–ca. 287 B.C.) tells us, "A stone which was found in the mines of Scaptehyle was similar in appearance to rotten wood, and would burn if olive oil was poured on it. When the oil had burnt away, the stone itself also would stop burning just as though it were unaffected by fire."[7] Translator D. E. Eichholz, following Robert H. S. Robertson, glosses this rather puzzling passage as a reference to the asbestiform mineral palygorskite (hydrated magnesium aluminum silicate hydroxide or mountain leather). The older interpretation, however, that of Nathaniel Fish Moore, is that Theophrastus refers here to asbestos at Thrace in the northern Aegean, probably in the modern region of Eski Kavala.[8]

In addition to the western sources, Chinese, Sinhalese, and Indian sources attest the use of asbestos in antiquity. The first definite identification of asbestos fabric in the Asian sources occurs in the work of Lih-tsze, writing at the end of the fifth century B.C. about fireproof cloth that was cleaned by exposure to fire. Alexander Wylie cites a number of other Chinese references to asbestos, dating from about 1,000 B.C. through the Ming dynasty, always as a material brought into China from foreign countries, usually India.[9] We also have a Mahawanzo Sinhalese reference to the famous Buddhist King Asoka (304–232 B.C.), who sent an asbestos towel from India as a gift to the king of Ceylon (modern Sri Lanka).[10] The fourteenth-century A.D. Chinese historical novel by Luo Guanzhong called the *San Kuo Chih Yen I* [*The Three Kingdoms*] mentions asbestos as having been known and used in China in the third century.[11] Ko Hung (Pao-p'u tzu) tells us in A.D. 320 that asbestos, along with gold and cinnabar, are among the most highly esteemed traditional Chinese medicines for "an eternal life."[12] Since the mineral seemed to have miraculous properties with respect to fire, it attracted similar beliefs regarding other apparent miracles.

Although the earliest western references to asbestos come from Greek sources and the name of the mineral is also Greek, Greek authors call the material *lithos amiantos* or *lithos Karystos*, after the names of the two most famous ancient asbestos quarries, on Cyprus and Euboea respectively.[13] As noted in chapter 1, Pausanias (after ca. A.D. 150) and Strabo (before ca. 64 B.C.–after ca. A.D. 21) mention incombustible lamp wicks, and there are references in Plutarch and Dioscorides as well. Dioscorides was the first of a number of compilers of materia medica and herbals to include asbestos in his compendium of natural wonders.

Because Pliny and the other ancient western commentators on asbestos play a significant role in the modern asbestos legal debate, it is worth setting forth in detail what these writings actually have to say about the mineral. Barry I. Castleman and others have incorrectly asserted that Pliny warned of the dangers of asbestos and that the slaves who worked the material wore dust masks. The elder Pliny was the first to call the mineral itself *asbestos*, as opposed to the *asbestinon* (fireproof) fabrics made from it mentioned by Varro the century before. There are a total of four references to asbestos, also called *amianthus*, or "live linen," in Pliny, and we shall examine each in turn. The first, in book I, is in his index to the *Historia Naturalium* [*Natural Histories*], in which he identifies the book and chapter in which his main discussion is to be found. True to form, the enthusiastic but unreliable Roman encyclopedist notes only one of his three later references, and not even the main discussion at that.[14] The next reference is in book XIX.iv, in which he discusses textiles and tells us that asbestos is a plant. I quote from Harris Rackham's translation in the Loeb edition:

> Chapter IV. Also a linen has now been invented that is incombustible. It is called "live" linen, and I have seen napkins made of it glowing on the hearth at banquets and burnt more brilliantly clean by the fire than they could be by being washed in water. This linen is used for making shrouds for royalty which keep the ashes of the corpse separate from the rest of the pyre. The plant grows in the deserts and sun-scorched regions of India where no rain falls, the haunts of deadly snakes, and it is habituated to living in burning heat; it is rarely found, and is difficult to weave into

> cloth because of its shortness; its colour is normally red but turns white by the action of fire. When any of it is found, it rivals the prices of exceptionally fine pearls. The Greek name for it is *asbestinon*, derived from its peculiar property. Anaxilaus states that if this linen is wrapped round a tree it can be felled without the blows being heard. Consequently this kind of linen holds the highest rank in the whole of the world.

In book XXXVI.xxxi, Pliny says, "Amiantus, which looks like alum, is quite indestructible by fire. It affords protection against all spells, especially those of the Magi."[15] Our last encounter with the mineral in Pliny is in book XXXVII.liv, where he says, "'Asbestos,' which is found in the mountains of Arcadia, has the colour of iron."

These passages, in which there is not so much as a hint of any risks to health (other than, perhaps, those of Persian witchcraft), are all Pliny has to say of asbestos.[16] They were summarized in the second half of the third century A.D. by Gaius Julius Solinus in his *Collectanea Rerum Memorabilium* and cited by dozens of later sources.[17] There is no mention in any of them of slaves, nor is there any reason to suppose, on the basis of ancient texts, that asbestos in western classical antiquity was ever mined or woven by enslaved laborers.[18] The bladder face masks worn by artisans over their faces in Roman times were used for the working of cinnabar, an ore of mercury, not asbestos, according to Pliny's book XXXIII.xli. Pliny does not say these artisans were slaves.

Before the eighteenth century, some historians and classicists were skeptical of Pliny's claim that asbestos could be woven into cloth, while others argued for its plausibility, noting the reappearance of the material in later texts such as that of Marco Polo.[19] Adrien Turnèbe (1512–65), a classicist who commented on Varro, among other ancient authors, was inclined to think the Romans knew the material.[20] Among those who doubted the feasibility of weaving the fiber at all was Johann Schild (1591–1667), who asserted flatly in 1662 that the thing simply could not be done.[21] A less radical doubter was Marcus Zuerius Boxhorn (1612–53), who thought anything might have been possible in the tropical wilds of India, but he and noted classicist Isaac Casaubon (1559–1614) were certain that the Romans never had any such fabric as asbestos.[22] It was not until, first, ancient Roman asbestos textile artifacts were found in Italy in 1633 and 1702 and then experiments with the material were performed between 1680 and 1750 with published results that this dispute was finally settled in Pliny's favor.[23] Eyewitness accounts by Johann Georg Keyssler (1693–1743) and James Edward Smith (1759–1828) of the asbestos shroud found outside the Porta Major in Rome in 1702 and thought to date no later than the time of Constantine lent additional credibility to Pliny's account.[24]

Archaeological evidence, earlier and later European writings, and Asian texts generally support and add details to Pliny's somewhat vague and erratic accounts of asbestos in western antiquity. Isidore of Charax, for example, writing in the first century B.C., describes trade that included asbestos cloth moving both east and west from India through the area that was later traversed by the Silk Road.[25] There is some archaeological evidence suggesting that asbestos may have been used in India/Parthia in the third century B.C. to make molds for casting shellac Buddha plaques.[26] If indeed asbestos fabric was brought to Rome from her archenemy Parthia, it would certainly have been expensive.

If, as the ancient authors imply, the mineral was mined, spun, and woven in Roman times only on the Greek island of Euboea, Amiantos on the northern side of the island of Cyprus, and in Parthian India, then it probably was not spun or woven by Roman slaves. Euboea and Cyprus became part of the Roman Empire in 78 and 22 B.C. respectively, but by this time asbestos was no longer mined in Euboea.[27] Plutarch (born before A.D. 50, died after 120) tells us in *De Oraculorum Defectu* that the asbestos resources of Karystos were played out before his time. If we are to believe the *Sung-shu*, written by Yueh Shen about A.D. 500 and discussing the century that had just passed, asbestos was woven in the Roman province of Syria, called Ta-ts'in in Chinese; the Chinese were not aware of any sources farther west at that time.[28]

Asbestos in the Middle Ages and the Renaissance

Asbestos appears early in Christian patristic writings after the classical period, where it retains the mythical quality we have seen in the Asian sources, although it is occasionally mentioned in practical terms. Pope Damasus I (305–84), for example, tells us in *In Silvestri Papa* that in the Emperor Constantine's time Saint Silvester, the first pope, ordered all the lamps in his baptistry to be fitted with asbestos wicks. There is a similar reference to amiantum in the *De Elia et Jejunio* of Saint Ambrose, Bishop of Milan, who died in 397. Asbestos was, as we have seen, already a popular material for the eternal lights of prosperous temples throughout the ancient world.

Another tradition, however, among Christian writers, appears first in Saint Augustine (354–430), the African Bishop of Hippo, who mentions asbestos three times in *The City of God*; but he seems confused about what it is, no doubt as a result of reading Pliny. He takes its Greek name literally and says in book XXI, chapter 5, "There is a stone found in Arcadia, and called asbestos, because once lit, it cannot be put out."[29] To him, the phenomenon of asbestos, however he misconstrued it, was evidence that the miracles of God's creation were beyond our understanding and beyond the power of skeptics to deny.[30] The fifth-century African compiler of Augustine's writing, Saint Eugyppius, seems to be laboring under the same misconception in his *Thesaurus*, as was Saint Bede (673–735) in the *Excerptiones* attributed to him. The same error from literal translation appears in Rabanus Maurus, Archbishop of Mainz (784?–856), in his

De Universo and *Excerptio de Arte Grammatica Prisciani*, and in Prudentius, Bishop of Troy (ca. 844–61), in his *De Praedestinatione*. Even Saint Isidore of Seville (d. 636), well known as an early mineralogist, says of asbestos that it is a stone "qui semel accensus nunquam extinguitur" [that cannot be extinguished once ignited]. Marbode of Rennes (1035–1123), clearly relying on Isidore's account in the *Etymologiae* at 16.4.4, says of asbestos in his verse treatise *De Lapidibus* [*Of Stones*], that it comes from Arcadia, is a wonder of nature, and cannot be extinguished once set on fire.[31] Pliny is obviously the source for the Arcadian location. Albertus Magnus's (1193?–1280) *Book of Minerals* relies heavily on Isidore and Marbode, and the author is just as confused about asbestos, or "abeston" as he styles it.[32]

There are similar passages in the patristic writings, such as the *Epitaphium* of Saint Eugenius, Archbishop of Toledo (d. 657), the *De Divisione Naturae* of Johannes Scotus Erigena (ca. 810–ca. 877), and the *De Divina Omnipotentia* of Saint Peter Damian (1007?–72).[33] Saint Gaudentius, who died in 410, was clearly a Pliny reader and cites his source on asbestos in a letter to Marcellus in which he compares the contemporary reverence for the mortal remains of martyrs with the separation of the sacred bones of royalty in their asbestos shrouds. Stephanus of Byzantium also refers to this practice in his *Ethnicorum*, written in the sixth century. Quoting from Hierocles, whose works are now lost, Stephanus says that the Brahmins of India wrap their important dead in asbestos so their ashes will remain separate from those of the other corpses on a funeral pyre.[34] These two sources at least correctly describe the use and one of the properties of the mineral. None of the western early medieval sources, however, write about asbestos with the confidence of personal familiarity. A reader has the unverifiable impression that none of them, from Damasus to Marbode, had ever actually seen either the mineral or the cloth they knew to have been made of it. Even in the sixteenth century, although most scientific authors gave reasonably accurate accounts of asbestos, Simeone Maiolo, Bishop of Volturara (1520–97) and Claude Saumais (1588–1653) were still propagating the error of Augustine and Isidore regarding the inextinguishable quality of the mineral.[35] In the Islamic world, however, the material seems to have been known and understood; Clare Browne notes that Abu Ubaid Al-Bekri (1040–94), a Moorish Spaniard, mentions the use of asbestos in North Africa for cordage and animal halters.[36]

The medieval evidence from the fourth century on seems, in fact, to suggest that, although there were asbestos artifacts and specimens in Europe at the time, the authors of lapidaries (books on minerals) did not connect these objects with the ancient accounts of asbestos because of the confusion about the supposedly inextinguishable flammability of the stone. This impression is reinforced by the dearth of passages in the patristic works in which asbestos is identified with amiantos/amiantus. Isidore and Bede write of both without connecting them, as Faustinus Arevalus noted in the late eighteenth century; Saint Jerome (died 419 or 420) mentions amiantus but not asbestos, as does the fifth-century author Blossius Aemilius Dracontius. When Rabanus Maurus and Anastasius the Librarian (ca. 810–ca. 878) use the root *amiant* they mean "never extinguished" in the context of sanctuary lamps, a usage that still seems to have been current in the time of Godefridus Ghiselbert (after 1659). There is no indication that any of these authors except Arevalus, who is, of course, a relatively modern commentator, recognize that amiantos and asbestos are the same mineral described in Pliny, Pausanias, and Strabo.[37] It may be this phenomenon to which asbestos industry chronicler W. E. Sinclair refers when he erroneously stated in 1959 that "records of asbestos in Europe were completely lost sight of for nearly eight hundred years."[38]

The general ignorance of asbestos during the Middle Ages apparently permitted a thriving trade in fraudulent relics and magical textiles such as the apocryphal tablecloth of Charlemagne (742?–814), thrown into the fire after dinner to amaze guests and win bets that it would not burn.[39] Considering the table manners of the period, it would no doubt have been a very practical household item, saving considerable labor in the royal laundry. We learn from Ahmad ibn Muhammad Ibn al-Faq ih al-Hamadh an i, called Ibn al-Fatiq, that in the tenth century Christian pilgrims to Jerusalem were sold small pieces of reddish-brown (ligniform) asbestos as pieces of the True Cross, its incombustibility supposedly proving its divine and magical properties.[40] The same "proof" of divine authenticity was used on a "Buddha garment" more than twenty feet long, presented to the Chinese Hou Wei dynasty king Kao Tsun (452–465), also called Wen C'en, by a king of Kashgar. Berthold Laufer tells us that this is only one of a number of incidents in Chinese history of testing asbestos by fire.[41] In this case, it was the Indian, rather than Middle Eastern, provenance that added plausibility to the object's claims for messianic associations. The tendency to add divine associations to asbestos's already marvelous qualities is clearly visible in both Asian and western traditions before the modern era.

The best-documented western case of asbestos relic fraud, however, is that of the cloth described by Leo Marsicanus, called Ostiensis, in his *Chronicon Casinense*, L.ii.c.33, written in the eleventh century. Some monks from Monte Cassino had gone on pilgrimage to Jerusalem and brought back a cloth sold to them as the original towel with which Jesus of Nazareth had washed the feet of his disciples. Again, the test of authenticity was the "relic's" incombustibility and capability of being cleansed by the action of fire. James Yates argues, on good evidence, that this artifact is the cloth seen in *Codex Casinense* 99 f.1, in which Monk John and Abbot Desiderius present Saint Benedict with the *Codex* while a novice kneels to wash the saint's left foot (see fig. 2.1).[42]

FIGURE 2.1 In this image from the Abbey of Monte Cassino in Italy, Leo kneels at the feet of Saint Benedict, washing his left foot. Textile historian James Yates argues that this cloth is the asbestos "relic" brought back from Jerusalem as the original cloth used by Jesus of Nazareth to wash the feet of his disciples.

From *Codex Casiensis*, reprinted in Herbert Bloch, *Monte Cassino in the Middle Ages* (Cambridge, Mass.: Harvard University Press, 1986), 3:1169.

Western knowledge of asbestos made a quantum leap in the mid-thirteenth century, when Marco Polo returned from Asia with a fairly detailed account of asbestos mining and fiber preparation in Chen-Chen. Significantly, his informant, Curficar, who had supervised the mining for the preceding three years, was from the Middle East—in William Marsden's translation, "a Turkoman"—confirming the earlier observation that the Chinese regarded asbestos as a foreign material associated with India and Syria.[43] Englishman John Speed's map of Tartary, published in 1626, included Polo's information about the location of asbestos in the east.[44] When Lodovicus Caelius Rhodiginus (1469–1525) wrote of asbestos a little more than two centuries after Polo, he also placed the source of the mineral in India.[45] Archibald Rose visited the Lolos, on the Chinese side of the border with India, about four hundred years later, in the first decade of the twentieth century, and found "an old chief in a long shapeless garment which was woven from asbestos strands," who "resisted all allurements of exchange."[46]

Sinologist Berthold Laufer recounts a number of Chinese and other tales about asbestos, including Asian speculations on its origin, of which perhaps the most colorful is that its source is a very large rat that lives on volcanoes, with silvery white hair ten feet long. Laufer associates this mythical beast with western notions of asbestos as the hair of the salamander, to which the equally mythical Prester John refers in the famous letter about his eastern kingdom, thought to have appeared in Europe in the late twelfth century.[47] Here again, the traditions conflate the unusual and, at the time, inexplicable fire resistance of asbestos cloth with far less credible tall stories about how such a wonder came to exist. Fire-resistive amphibian legends are the source of the logo image of the International Association of Heat and Frost Insulators and Asbestos Workers: a salamander over flames.[48]

However fanciful, the efforts of early scientists to classify asbestos as either a plant, as Pliny does, or as the hair of an animal are rational in origin: all other textile materials before the twentieth century were in fact the products of either animals or plants. As Stephen Pyne reminds us when he tells us that "fuels were alive," any organic product is by definition combustible; and it was thus completely counterintuitive, to the inquiring premodern mind, that there should be a stone from which it was possible to make fabric and that there should be fabric that would not burn.[49] Charles Bonnet even proposed in the eighteenth century that asbestos is the missing link between the mineral and vegetable kingdoms in God's great and supposedly immutable chain of being.[50] Asbestos continued to be a source of wonder and curiosity to scientists well into the nineteenth century, as we shall see.

Asbestos in the Early Modern Period

In the late fifteenth and early sixteenth centuries the investigation of asbestos and its properties began to take on some of the formal trappings of modern science as received classical knowledge, empirical traditions, and what Aaron J. Ihde calls the "alchemical and the technological heritages" began to converge into the discipline now known to us as chemistry.[51] Accounts of the mineral are generally less colorful and more mundane, and there is greater reliance on eyewitness data. The chemists and mineralogists of this period were particularly interested in making distinctions between various minerals of similar appearance by using sensory data other than sight as well as chemical testing. Travelers also published reports of asbestos in use in Europe and the east.

German metallurgist Georgius Agricola (1494–1555) was one of these practical and empirical authors, who, like his contemporary Pietro Andrea Mattioli (1500–77), was interested in the question of how to distinguish asbestos from alum, which, as Pliny remarked, the mineral resembles. Much of Mattioli's discussion of *lapis amiantus* is devoted to this issue and to a recapitulation of Pliny's account.[52] In his *De Natura Fossilium* of 1546, Agricola summarized existing knowledge of asbestos, adding that Zoroaster called the material *bostrychites* (braided.) Agricola, who had clearly seen it with his own eyes, explains that skeins of asbestos were usually sold braided. He notes that the first-century B.C. author Quintus Claudius Quadrigarius mentioned the use of asbestos for fireproofing but confused it with alum because, the German says, "it has a fracture similar to alum." The confusion was particularly understandable given that both materials have traditionally been used for fireproofing. Agricola, Mattioli, and later authors Ole Worm (1588–1654) and Athanasius Kircher (1602–80) recommend tasting asbestos for its characteristic "slightly astringent" quality, which serves, in a period of chemical inquiry with little in the way of precision instruments, to distinguish it from its look-alike, so-called feather alum, or *alum scissile*, which has a much more acrid taste.[53]

The learned disquisitions of chemical authors on asbestos were elaborately spoofed during Agricola's and Mattioli's lifetimes by François Rabelais (ca. 1490–1553?), who devotes an entire chapter of *Gargantua and Pantagruel* to an absurd description of "pantagruelion," a fireproof mineral. "Don't talk to me about feather alum," he exhorts in chapter 52, adding a distinction to asbestos's literary pedigree not to be equaled until Victor Hugo in the 1860s included the mineral in a Gothic novel set in Iceland. In the twentieth century, the movie version of L. Frank Baum's *The Wizard of Oz* gave the Wicked Witch of the West an asbestos broom, extending its artistic cachet—and mythical qualities—to the silver screen.[54]

Worm and Kircher both owned specimens of asbestos and both, like some of the Chinese authors, were interested in its chemical, medicinal, and magical properties. Worm thought it was useful in childbirth, for sores, and for diaper rash. Kircher, who is probably best known for his theory that the earth was hollow and had another inhabited surface inside it, used asbestos to make paper and thought the stone valuable for use against enchantments, whether Persian in origin or not.[55] Scientific writers as late as the eighteenth century still regarded the material as virtually miraculous. Sir James Edward Smith, for example, saw the famous asbestos shroud at Rome and reported the episode in awestruck terms: "We were shewn, by express desire, the winding-sheet of Asbestos. It is coarsely spun, as soft and pliant as silk. Our guide set fire to one corner of it, and the very same part burnt repeatedly, with great rapidity and brightness, without being at all injured. I have no conception what the flame could feed on. Its brightness seemed to indicate nitre."[56]

The longest scientific treatise on asbestos from this period is that of Matthias Tiling (1634–85), a German physician who, like his contemporaries, quotes extensively from earlier authors, including information about the usefulness of the material as a talisman against witchcraft, but also recounts his own experiments with "amianthus stone." Among these are medications much like those of Worm, for which Tiling provides formulas: an ointment for diaper rash and sores of the extremities and a remedy for "album fluxum" (leucorrhea) in women.[57] Giovanni Giustino Ciampini (1633–98), Nicholas Mahudel (1673–1747), and Franz Ernst Bruckmann (1697–1753) also wrote scientific essays on asbestos, the last a mineralogist who reportedly experimented with printing on asbestos paper, as did Edward Lloyd in 1684.[58]

Travelers' accounts of asbestos in the fifteenth and sixteenth centuries make it quite clear that William E. Sinclair is in error regarding the material's having fallen into entire desuetude in Europe after 800. Browne reports the use of asbestos in armor and festive dress in thirteenth- and fourteenth-century France and Wales.[59] Juan Luis Vives (1492–1540) saw both asbestos lamp wicks and napkins in Paris, and Pietro della Valle (1586–1652) was given some asbestos cloth at Larnaca on Cyprus in the following century, although he was also told it was no longer made there and remarked, "At this day none knows how to make the Cloth, or to spin the matter; although a whitish matter like Cotton is clearly seen to issue out of the stone, not uncapable of being spun." Joseph Pitton de Tournefort (1656–1708) saw asbestos being softened with olive oil and spun with flax; but according to Charles Sigisbert Sonnini (1751–1812), the mines in northern Cyprus were not reopened until the market for asbestos improved in the late eighteenth century.[60]

Asbestos and the Scientists in Britain and America, 1700–1850

By Pitton de Tournefort's time, asbestos was well enough known in Europe to have attracted the attention of the Royal Society in Britain. The author of "Inquiries for Turky" asked in the *Philosophical Transactions* in 1665, "What store of Amianthus there is in Cyprus; and how they work it?" A few years later the *Transactions* published an account of asbestos paper, fabric, and candlewick from a Venetian source.[61] A fragment of Chinese "linen" cloth was reportedly shown to the society in 1676, and a series of articles and letters about the mineral appeared in the *Philosophical Transactions* between 1680 and 1741. Robert Plot, for example, published an article summarizing previous knowledge of asbestos in 1685, called "A Discourse Concerning the Incombustible Cloth," following up a series of experiments performed at a society meeting. In 1712, Patrick Blair wrote to Dr. Hans Sloane, who later bought Benjamin Franklin's asbestos purse, about a house he had seen in the Scottish highlands in which asbestos had been used as the building stone. Systematic scientific investigation was soon under way in Britain and Europe and, a little later, in the American colonies. The mineral was included in results of experiments with a burning mirror in 1718, a list of minerals found in Sweden in 1713, and an account of the natural history of Greenland by Hans Egede (1686–1758) in 1741.[62]

By the end of the eighteenth century, asbestos was well integrated into contemporary investigations of chemistry and mineralogy. When Swedish scientist Torbern Bergman wrote about asbestos in 1782, he had performed a number of experiments on the material, using specimens from different parts of Sweden and the Tyrol to determine its properties and composition, which he declared to be "siliceous, magnesian, calcareous, and argillaceous earth, with some admixture of ferruginous matter." He reports the results of "dephlogisticating" the material (driving off any combustible components) and discusses the difficulty of finding any practical use outside the sciences for asbestos: "as yet it has never been otherwise considered than as a curious phenomenon of physics." He goes on to report his experiments with incombustible wicks, which, he tells us, require frequent cleaning to prevent clogging and ultimate extinguishing of the flame. As to its utility for other purposes, Bergman is skeptical:

> The asbesti have been hitherto applied to little or no use. Formerly, indeed, cloths made of the softest kinds were employed to wrap the bodies of the dead, that, by its qualities of resisting fire, their ashes might be preserved. But on the abolition of funeral piles [*sic*], the utility of the asbestos ceased. And as to its being calculated for garments for the living, the continual and intolerable irritation of its harsh and short fibres would render it certainly not very desirable.

He goes on to comment on the impracticality of manufacturing asbestos writing paper, which he describes as "both brittle and absorbent," neither quality being advantageous in a surface to which to apply ink.[63]

Other uses, however, nearly all of them scientific, were extending asbestos's visibility beyond the cabinet of curiosities. Scientific demand accounts for the change in the international market for Cypriote asbestos noted by Sonnini in the 1790s. Between 1750 and 1850, twenty-four articles in the *Philosophical Transactions of the Royal Society* mention the mineral; of these, only four were mineralogical in character and one archaeological. The remainder, authored by scientific luminaries such as Joseph Banks, Humphry Davy, Andrew

Ure, and Michael Faraday, refer to uses of asbestos in experiments and observations in chemistry, physics, astronomy, metallurgy, and biology, usually as a filtration material.[64] When Benjamin Franklin sold a purse of the material to Sir Hans Sloane of the Royal Society in 1724, it was worth enough to enable him to continue his education in London. Franklin remained interested in the mineral and had pieces of it in his collection a quarter of a century later; Alexander Hamilton collected notes on it in his artillery paybook.[65] Swedish scientist Pehr (Peter) Kalm (1715–79) reports that, according to Franklin, the "stone . . . on account of its indestructibility in fire is used in New England for making smelting furnaces and forges" as well as fireplaces of other kinds.[66] Mineralogists were hard at work as well, trying to establish "species" of rocks and minerals analogous to the classification of living organisms by Linnaeus and others. These efforts had limited success, but much was learned about the chemical and geological aspects of asbestos through the researches of eighteenth- and nineteenth-century mineralogists and geologists such as Giacinto Gimma (1668–1735), Emanuel Mendes da Costa (1717–91), Axel Fredrik Cronstedt (1722–65), Johann Reinhold Forster (1729–98), Samuel Robinson (1783–1826), and Parker Cleaveland (1780–1858).[67]

Some traditional uses of asbestos continued through the nineteenth century and into the twentieth, including incombustible wicks for oil, alcohol, spirit, kerosene, and vapor lamps; lime (calcium) lighting; and, later, wick pads for burning petroleum under boilers.[68] Nathan Rosenberg, citing Charles Coleman, reports that the Welsbach asbestos mantle for gas jets, which produced six times as much candle power as an unmantled jet for the same flow of gas, added fifty years to the commercial life of gas lighting after the invention of electric light.[69] Reportedly, asbestos also remained in use for purposes of religious fraud in some parts of the world: Eusebe Salverte reported in 1845 that plasters of asbestos were applied to the feet of those who "commanded public veneration by walking over burning coals."[70] Because it was known to be a relatively rare and economically valuable mineral, asbestos was routinely included in eighteenth- and nineteenth-century surveys of mineral resources in the United States and elsewhere.[71]

In the mid-nineteenth-century United States, asbestos appeared regularly in the pages of *Scientific American*, *Science*, and other technical journals after 1845 as a standard chemical supply item for scientific uses. The material remained too expensive for any but technical and highly specialized applications of small quantities until the second half of the nineteenth century, although it was in limited use as an insulator for safes in 1835 and in boiler packing by 1849.[72]

Four qualities of asbestos made it interesting to the scientific and technologically inquiring minds of American inventors of the mid-nineteenth century: the material resisted acid, it had a very low thermal conductivity, it made a fine and very reliable filter, and it was impervious to fire. Experiments continued in the new century; and new applications of the material were attempted, including a type of dental packing developed in 1849 and a filling in 1852 that exploited the thermal nonconductivity of asbestos to protect sensitive dental nerve endings.[73] Through the 1840s and 1850s, asbestos was advertised mainly in scientific periodicals, including *Scientific American*, which regularly accepted advertising from a Dr. Lewis Feuchtwanger, of Maiden Lane in New York City, mail-order and retail purveyor of "Rare chemicals, metals, soluble glass, oxyds [*sic*], uranium, cobalt . . . asbestos, French chalk, insect powder," and other chemical products.[74] Editorial matter in the magazine commented frequently on the growth of uses for asbestos in the second half of the nineteenth century, and patents for new products containing asbestos were regularly posted in the "New Inventions" and "List of Patent Claims" columns. *Manufacturer and Builder* also reported on developments in the field of asbestos use.[75] Dozens of articles appeared reporting the results of experiments in which asbestos played an important role as a filter, including those of J. Terreil and Louis Pasteur in the 1850s and 1860s disconfirming the theory of spontaneous generation. Fungi and other organisms grew in cultures in contact with unfiltered air; but when the air was filtered through asbestos or any other very fine filter, no organisms appeared in the medium. Thus, it could not be the case that these organisms were spontaneously generated from the air. Rather, they must have grown from microscopic but filterable particles in the air.[76]

Other reports of scientific and engineering experiments using asbestos filters included testing the air in a wallpapered room for arsenic, making a compound used in carburetion called porous carbon, filtering alkaline solutions, and making "permanganic acid in aqueous solution." In some of these applications, the acid resistance of asbestos made it one of a limited number of possible choices.[77] Scientists were interested in other uses for the material as well: Camille Faure invented a type of battery, reported in *Science* in 1888, in which "each plate is surrounded by a sheet of prepared asbestos"; and experiments were made in the 1890s with asbestos hot-air balloons that would not burn as did their silk counterparts. Asbestos was incorporated into a number of types of scientific gear after 1870, finding applications as a lightweight hut-building material in the Antarctic in 1902–4, in the construction of chemical laboratory and herbarium walls and tables, as washers in oxygen breathing apparatus for the Mount Everest expedition in 1922, as a fire and heatproof box lining for photographic negatives on the Hamilton Rice expedition to the Amazon in 1924–25, and as oil pump and tank insulation on the aircraft used by Admiral Richard Byrd in his Antarctic exploration in the late 1920s.[78] Sheet asbestos and asbestos packings were available by the late 1880s from H. W. Johns and a few other manufacturers and were employed for insulation of scientific equipment as well as in building and thermal insulation of boilers and other heating devices. Vulcanized asbestos gaskets were also being manufactured during this period.[79]

Paper, which as we have seen was an experimental use for asbestos as early as the sixteenth century, continued to receive attention from nineteenth-century scientists and inventors in Europe and the United States, who would usually blend the mineral with a vegetable fiber such as cotton or jute for applications such as "incombustible account books." George Smillie of New York City attempted to create a green ink based on asbestos in 1863, apparently hoping to cash in (as it were) on what proved to be a short-lived commercial interest in fire-resistive currency and bank notes.[80] In 1856, Edward Lindner patented a gun breech containing asbestos for "the purpose of cleaning the rubbing surface of the breech in its motion upwards"; and other inventors devised asbestos-containing lubricating compounds for journal boxes in the 1860s and 1870s.[81]

Although the material was known mainly to scientists and inventors before 1860, a protoindustry began to form in midcentury around asbestos's resistance to fire and its low thermal conductivity. The first mass-produced asbestos-containing product used in the United States appears to have been gas fireplaces introduced by the British firm Bachoffer & DeFries in the early 1850s. This development inspired derision in devotées of high culture, who considered the "artificial" fire vulgar and absurd. One such writer even questioned whether a cat would deign to sleep before such an affront to the secular religion of the hearth. In both Europe and the United States, asbestos was in use by 1852 in the fireboxes of gas cookstoves and, by the mid-1860s, as burner padding under chemical and culinary glassware to slow down the process of heating, a procedure useful in fractional distillation and for cooking ingredients that had a tendency to stick and burn. Mats of this type were available in the United States until the mid-1970s.[82]

In Italy, Count Aldini had begun making protective clothing—what is now called a proximity suit—of asbestos on wire gauze in the 1830s for firefighters in Geneva and Florence.[83] In Austria, asbestos proximity suits of the Italian type were available to firefighters by 1864, when a reader of *Scientific American* expressed surprise that a product so obviously practical was not available in the United States.[84] By 1890 firefighters in Paris and London were wearing asbestos clothing and using asbestos salvage blankets and bags.[85] Some brands of metal safes in the 1860s had asbestos insulation. The mineral's fire-resistive qualities began in the 1870s to figure in popular fiction, such as Edward Everett Hale's science fiction story "Life in the Brick Moon," in which asbestos is proposed as the insulating material for a kind of nineteenth-century spacecraft, and in chapter 72 of R. D. Blackmore's serialized novel *Alice Lorraine*, where it appears as a protective covering for a cache of gems.[86]

It was in the 1860s and 1870s, in fact, that the modern uses we associate with asbestos began to emerge. Asbestos boiler packing, for example, was reportedly patented in the United States by 1828 but was still too expensive to gain much ground in either the land-based steam or maritime markets until later in the century. In 1878, *Manufacturer and Builder* reported that asbestos cost twice as much as mineral wool and that the material's sticker-shock aspect was the principal obstacle to its wider use as insulation. Efforts were made to reduce the cost of using asbestos by mixing it with less expensive packing materials.[87] For specialized insulating purposes, such as "the joints of apparatuses exposed to hot acid vapors," asbestos was already the recommended material, mixed with silicate of soda, then called "water glass."[88] The price was just beginning to fall in the 1870s when both *Scribner's Monthly* and the *Galaxy* praised its qualities as an innovative and relatively affordable insulating material.[89] Even in 1897, however, price resistance remained a significant limitation on the marketing of asbestos and its products.[90] John F. Springer, writing in 1912 about asbestos's transition from an expensive curiosity to an article of commerce, noted that the price of Canadian asbestos, mined by the old methods in 1890, had been a prohibitive $128 per ton and that by 1904, mainly because of more modern mining methods and improved transportation to markets, the price had fallen to a still fairly expensive $30 per ton.[91]

Asbestos roofing, felting, and other insulating products were prominently featured at American industrial exhibitions in the late nineteenth century. The Cincinnati Exposition of 1873, for example, which *Manufacturer and Builder* called "the best exposition of the industrial arts ever held in this country," won praise for its state-of-the-art steam-power system, "the pipes being covered with an asbestos preparation known as the Chalmer-Spence Patent Non-conductor."[92] Henry W. Johns, who invented new methods of working the mineral at an acceptable cost, attracted significant commercial attention at the American Institute Fair of 1875 and then won first prize at the Centennial Exhibition the following year for his line of asbestos products, including boiler packing, roofing, cement, paint, millboard, cloth, thread, and paper.

The enthusiasm for the cost-effectiveness of asbestos boiler insulation expressed in scientific and engineering publications had become positively effusive by this time; of the 1875 exhibition, *Manufacturer and Builder* said that the Johns products formed "a magnificent display . . . which cannot be recommended too highly."[93] Johns had been an advertiser in *Manufacturer and Builder* since 1871 and in *Scientific American* since 1868 (the year in which he patented roofing fabric), which no doubt explained some of the ardor of the editors' support; but in fact the market for asbestos products was growing mainly because they performed as advertised.[94] The material did not deteriorate in use, did not catch fire, was not thermally conductive, was lightweight relative to metal and ceramic materials, did not stink when hot or harbor vermin the way hair felt did, and insulated so effectively as to permit impressive cost savings in fuel. *Scientific American*, writing retrospectively of the beginnings of asbestos use in 1946, said:

> Early in the development of steam engineering, it was recognized that radiation, convection, and conduction from boilers and hot lines were robbing steam engines of much of their efficiency. Insulating materials of various types were experimented with, most of them of an organic base—vegetable or animal matter—and most of them were subject to decomposition at around 300 degrees Fahrenheit. Later, plaster of Paris mixed with sponge was tried—the small entrapped dead air cells of the porous sponge being a fairly good insulating substance—but the plaster rusted the pipes and boilers. Other insulating materials were also used and found unsatisfactory in that they shrank, charred, would not resist vibration, or otherwise deteriorated in use.[95]

A hospital for the insane in Middletown, New York, reported in 1879 saving two hundred tons of coal a year "by the use of asbestos pipe coverings."[96] In August 1876, *Manufacturer and Builder* noted, "Not many years ago asbestos was a rare product, found only in mineralogical cabinets." Within a few months, the editors wrote breathlessly that

> asbestos steam packing, which has been recently improved and perfected, and having been thoroughly tested, is claimed as being the best, cheapest, and by far the most durable packing in use for piston-rods, valve-stems, throttle-valves, pumps, and stuffing boxes of high and low pressure engines, etc. The long and severe tests to which this packing has been subjected, and the surprising results attained—in one instance having been used, without removal, on an ocean steamer which sailed over 90,000 miles, and in another on a locomotive which ran over 50,000 miles—and as it is indestructible by acids, long exposure to dampness or any degree of heat, and also possesses the peculiar property of being a natural lubricator which does not heat or waste perceptibly, the fact of its absolute economy over any other article for similar purposes, aside from the great saving by avoiding stoppages, is self-evident.[97]

The Federal Bureau of Steam Engineering tested 85 percent magnesia steam pipe insulation in 1887, a product consisting, not surprisingly, of 85 percent magnesia plus 10 percent asbestos, with the remainder consisting of impurities of various kinds, usually organic fibers. The product was found "superior to hair felt" as well as to mineral wool and "infusorial [diatomaceous] earth" as an insulation material. By 1905, the Asbestos and Magnesia Manufacturing Company of

Philadelphia could plausibly assert that "since that day it has covered perhaps ninety per cent. of the steam surfaces, on land and sea, owned by the United States Government, and approximately seventy-five per cent. of all the high pressure steam-carrying surfaces in the United States."[98] So general was the acceptance of asbestos insulation among the technical elite of the late nineteenth century that when Robert E. Peary set out for Greenland in 1891, the alcohol stoves he took with him had asbestos-jacketed boilers.[99]

Asbestos gained ground at the end of the nineteenth century as an electrical insulator, an application in which it was to have a long career in the twentieth century. "Vulcabeston," an asbestos-rubber compound later called "ebonized asbestos," was available by 1888 and was well regarded by electrical engineers, who used it for "dynamos, motors, arc lamps, converters, street-car controllers, switches, rheostats, thermostats, and many other forms of electrical apparatus" as well as in "electrothermic apparatus" such as car heaters and heating pads.[100] Asbestos-rubber switchboards were in common use in the early twentieth century.

In the same period, successful experiments were performed with other uses for asbestos, including some that resulted in patents. William Peters patented an asbestos-containing fire brick in 1862, and A. Pelletier included the material in a kind of purportedly fire-resistive stucco, patented in 1868. Arza Lyon of Cincinnati, a city rarely behindhand in nineteenth-century American industrialization efforts, patented a "Composition for Building" consisting of "a mixture of asbestos, hydraulic cement, coal ashes, air-slacked lime, and plaster of Paris, mixed with water, molded, applied, and then saturated with water-glass."[101]

It was as a roofing material, however, that asbestos had not only a substantial advantage from the perspective of fire protection but from that of cost as well. The competitive products—slate, ceramic tile, and tin—were all more expensive than asbestos, and the first two could not be applied to existing structures that had not been designed to carry their weight. Tin roofs were fire resistive but costly and were, of course, thermally and electrically conductive as well as subject to oxidation and corrosion. After the Chicago and Boston fires of 1871 and 1872, the combustibility of roof coverings and insulation received greater attention from architects and builders, with some incorporating asbestos into both roofing materials and heat insulation for the iron and steel beams that had shown such a distressing tendency to buckle when exposed to the high temperatures of urban conflagrations.[102] Built-up roofing containing asbestos felt was introduced in 1885; it was, however, initially expensive and labor-intensive to install.

Keasbey & Mattison had a reportedly durable "asbestos-metallic roofing" by 1902 for which the company claimed that "notwithstanding its naturally enhanced price, is the cheapest material to use for permanent buildings."

Johns-Manville, aware of the price resistance of the market, called its product "the cheapest per year" roofing. A patent for asbestos-cement, which consists of Portland cement mixed with asbestos fibers, was issued to L. Hatschek in 1904, and uses were quickly found for the material, including as a roofing material.[103] Asbestos-cement shingle roofs, available by 1906, cost more than asphalt shingles (about twenty dollars more per roof in 1923) but less than slate or ceramic-tile roofs, with comparable fire-resistive qualities. They required care in installation because, like ceramic and slate, they were breakable; but they did not split at the corners like wood and required very little maintenance once installed. Unlike ceramic tile, the material could be easily cut with a saw. The Asbestos Shingle, Slate & Sheathing Company of Ambler, Pennsylvania, pointed out as well that the product did "not crack or exfoliate when exposed to fire, as natural slates do."[104]

The indefatigable Henry W. Johns had developed by 1874 an asbestos paint that within a few years acquired a reputation for slowing down the rate of combustion of wood and preventing oxidation of metals such as iron. Johns took out a full-page advertisement for his paints and roofing in the *American Missionary* in 1878, citing the cost savings as a sales incentive. Although the paint actually cost more than competitive products of the period, Johns's sales technique must have been effective because by 1879 the U.S. Capitol and other federal buildings in Washington, D.C., were entirely coated with Johns's asbestos paint.[105] The famous fire-safety engineer and firefighter Eyre Massey Shaw performed tests in London on this paint and found that it significantly retarded the development of flame in wooden structures.[106] The results of such experiments are depicted in a series of drawings in the 1888 *Asbest und Feuerschutz* (see fig. 2.2), which probably exaggerates the fire-resistive effect.[107]

By 1880, asbestos was recommended by some experts in other building applications as well. James Jackson Jarves, for example, writing in *Scribner's Monthly* in 1879, warmly endorsed asbestos millboard as a wall covering for art museums, arguing,

> This substance possesses special advantages as a background for pictures and other works of art. Even if soaked with kerosene it will not burn, and it is virtually indestructible in any common flame. Being an atmospheric non-conductor it serves to keep a room warmer in winter and cooler in summer, thus helping equalize the temperature, an important object in a museum. As it is a purely mineral substance, it does not harbor insects or generate noxious odors. Its natural color is a soft, neutral tint, very favorable itself for art-objects; but it can be colored with warm tints if required. An absorbent and not a reflector of light, as are most wall-papers, it does not fatigue the eye or dazzle it by contrast with the object placed against it.[108]

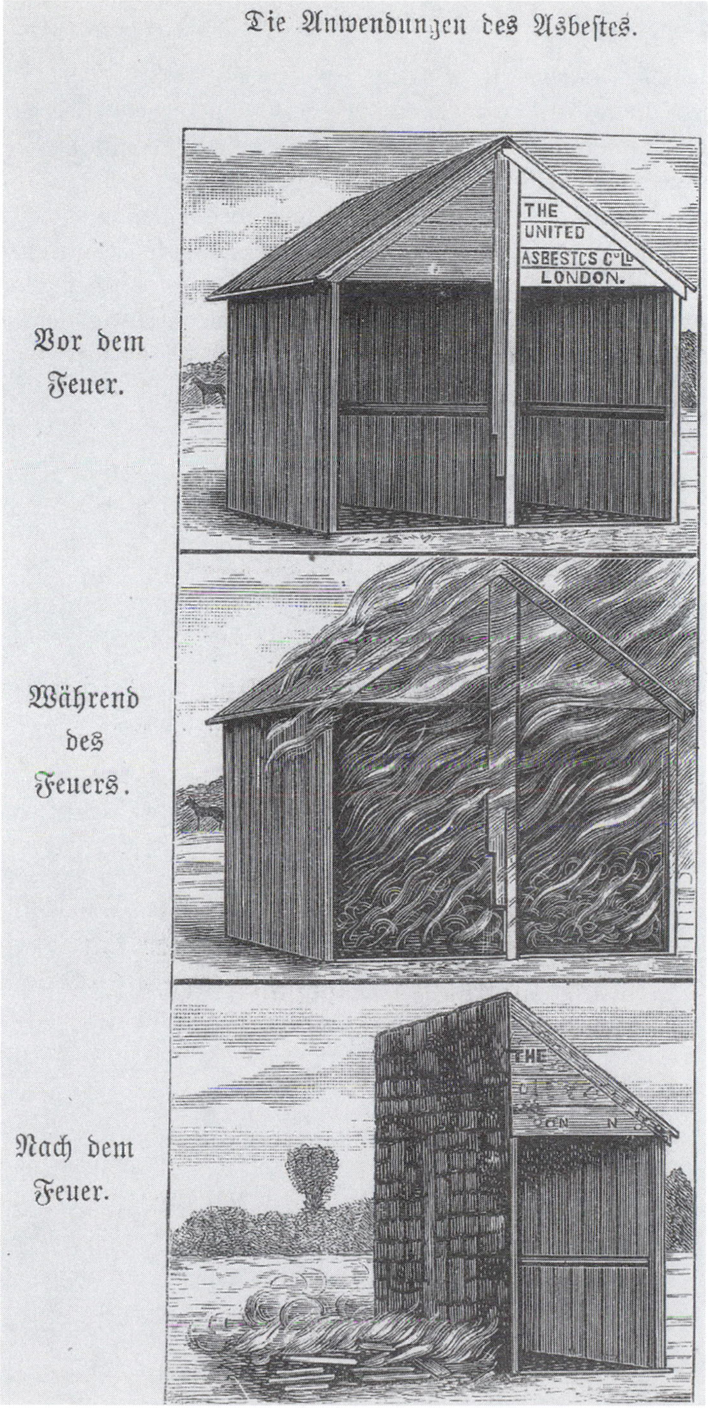

Die Anwendungen des Asbestes.

Vor dem Feuer.

THE
UNITED
ASBESTCS C° Ld
LONDON.

Während des Feuers.

Nach dem Feuer.

FIGURE 2.2 German illustration of a British test of asbestos paint, 1886. The exposed surface of the "United Asbestos" side of the shed is visibly charred in the third frame but did not ignite, according to Venerand.

Wolfgang Venerand, *Asbest und Feuerschutz* (Vienna: Hartleben, 1886), 108.

Scientific experiments by safety professionals involving the controlled burning of fire-resistive structures using asbestos building materials had begun by 1891, and the mineral had begun to attract the attention of fire insurance underwriters.[109] In the last third of the nineteenth century, asbestos was a mineral with an exotic past and a modern, cutting-edge reputation among scientists and engineers, the industrial era's analogue of a space-age material. It was a material still waiting in the economic wings of the increasingly dense built environment; but fire, one of nature's most powerful forces, was about to drive asbestos literally onto the stage of history.

3

The Rise of the Asbestos Curtain

Between 1797 and 1897, more than 10,000 persons worldwide are known to have perished in theater fires, nearly all of which started on the stage and then engulfed the audience in smoke and toxic gases.[1] This figure does not include small theaters in out-of-the-way places, most of which had no reliable records of such episodes, and no authorities to which to report them, nor does it include any public-assembly occupancies other than theaters, such as hotels, churches, and cabarets.

The appalling loss of life in nineteenth-century theaters created a demand both for life-safety technologies to prevent disasters such as the Richmond, Virginia, fire of 1811 and the Vienna Ring Theater fire of 1881 and for building codes to enforce their mandatory use. The period between the first quarter of the nineteenth century and the first decade of the twentieth was characterized by commercial and official experimentation with methods to prevent stage fires from spreading to the auditorium, beginning with the installation of iron proscenium curtains and rooftop smoke vents in some European and British theaters. By the time Edwin O. Sachs was investigating the merits of solid iron, iron wire, and asbestos curtains for making the stage and auditorium into separate fire areas in 1898, asbestos was already becoming the preferred medium for such curtains and was by 1909 explicitly recommended or required in the building codes of most large cities in the United States.

At the same period, other types of fires, particularly those in public-assembly occupancies in which multiple fatalities were common, gradually convinced civic authorities, fire-safety professionals, engineers, architects, and insurance underwriters that the increasingly urbanized environment in Europe, Britain, and the United States needed stricter regulation if fire mortality and economic loss were to be reduced to acceptable levels. This community of experts, like other technical occupations, became professionalized in western industrial nations in the early twentieth century, with its own international organizations, meetings, and methods of arriving at consensus regarding standards of safety.

Tickets to the Tomb: Theater Fires in the Nineteenth Century

Fire narratives, a few of which I shall summarize later in this chapter, make very depressing reading. Century after century, they are gruesome tales of combustible materials exposed to sources of ignition in structures containing too many persons, from which safe exits free of smoke and flames are missing, inadequate, inoperable, locked, or invisible under fire conditions. It is and has always been the responsibility of fire-safety professionals to read and study these narratives for the lessons they can teach; a fire-safety engineer recently remarked that the *Life Safety Code* should be treated as a sacred text, since every paragraph in it required the sacrifice of one or more human lives. As we shall see, the lessons of fire safety were all learned the hard way; and it is part of their heartbreaking character that so many of them are grimly repetitive, requiring that we learn them again and again at the cost of additional human lives. William Paul Gerhard, for example, might have been writing of the Station nightclub fire of February 2003 in West Warwick, Rhode Island, when he said in 1896, "The use of fireworks in spectacular plays calculated to attract the public; the use of red fires, torches, Roman candles, the firing of rockets,—are dangerous sources of fire, because a spark often lodges amidst the scenery." Or, he might have added, in the non-code combustible ceiling tiles. Of the 1,108 theater fires worldwide between 1797 and 1897 documented by Sachs in 1898, thirty-one were started by fireworks on the stage.[2]

As fire-safety professionals have come to understand it, fire safety is a complex system, requiring many layers of awareness and implementation of risk reduction. Sara Wermiel has shown that the elements of fire safety in building construction began to be studied systematically in the nineteenth century and to be embodied in building laws, first of cities and then of larger governmental units: states and provinces and ultimately, in places such as Britain and Canada, the national government. Most of the building codes currently incorporated into municipal and state law in the United States are based on one or more of the model codes developed by various communities of experts after 1896, such as the National Board of Fire Underwriters, the National Fire Protection Association, and the International Conference of Building Officials.[3]

The purpose both of the model and of the enacted codes, in the case of fire safety, was to increase the odds against a fire's ignition in the first place and then to increase the odds against its spreading, destroying property, and, most important, ending lives. The system of safety that is codified in, for example, the *Fire Protection Handbook* and its companion volumes the *Life Safety Code* and

Life Safety Code Handbook, is designed to stack the complex deck of the built environment in favor of human survival without injury in the event that preventive measures fail.[4] Numerous redundancies and fail-safes in codes and their enforcement improve the chances that even if some components of the system do not function as planned, others will. For example, in the Station fire of 2003, if fire-resistive ceiling panels had been used, the fireworks could not have ignited them. If there had been compliance with safety regulations regarding pyrotechnics with a proximate audience, there probably (but not certainly) would have been no fire to ignite the ceiling. If the stage had been protected by a safety curtain as a large theater's stage would have been (nightclubs without proscenium stages do not have such curtains), the smoke and flames could not have reached the audience. If exits had been more visible or more numerous, more of the audience would have left the building alive. One can proliferate these counterfactual conditionals almost indefinitely, but the responsibility of the fire-safety professional is to provide architects and building owners with proven methods of safeguarding the built environment against as many preventable dangers as are known so that even if several factors go wrong at once, the humans in the burning built environment still have a fighting chance for life.

Considering these factors mathematically, let us suppose that we are considering ten safety factors in a given environment, a relatively small number in the kinds of complex urban environments we will soon be examining. Even without taking Murphy's Law into account, some of these elements will fail in unforeseeable ways by chance, malevolent design, or human error: for example, the sprinkler system will have been turned off to fix a leak, a fire door will have been "temporarily" propped open or "temporarily" locked, a terrorist will fly an airplane into the building. If the success of safety factors were the heads sides of coins, the chances of their all failing at once (landing tails up) are 1,024 to 1. If even one of these factors fails, our chances of system failure rise to 512 to 1; if two of them do, they are 256 to 1. By the time more than half of our safety components have failed or are found to be missing, our chance of disaster has reached 1 in 32. Building codes and those who enforce them try to ensure that as many of these metaphorical safety coins as possible land heads up.[5]

Asbestos, first in theater curtains and motion-picture projection booths, then in insulation and roofing, began to appear in municipal building codes in the late nineteenth and early twentieth centuries as part of this professionally designed system of fire safety, a safety component built into the structures and spaces people lived, worked, and played in, because the experience of urban life without this system of protection had convinced us we could not continue to rely on the enlightened self-interest of property owners and managers. Nowhere was this more obvious at the end of the nineteenth century than in theaters.

Drury Lane in London is justly famous for its long history of theaters, including those designated Theatres Royal, of which there have been a noble succession since Christopher Wren's edifice was built after the Great Fire of 1666. Most have been destroyed by fire; Wren's more fortunate building was simply condemned and razed in 1791.[6] Henry Holland's new Theatre Royal of 1794 was a gigantic brick structure that held 3,600 people in five wooden tiers of seats above the orchestra. Theater historian Donald C. Mullin clearly does not regard Holland's supposedly fire-resistive design as a success:

> To prevent the building from flaring up at the first overturned candle, a great cast iron fire curtain was installed, and large tanks filled with water were placed in the attics for emergency use. Naturally, the house burned down, for by 1809 the iron curtain had rusted into uselessness and had been removed, and the water tanks had long since sprung leaks and run dry. It was later commented that if the house had not burned, it would have tumbled down because of the shoddy construction.[7]

Holland had good reason for concern, however. Theater ownership in his century and the century afterward was a classic example of the triumph of hope over experience. The famous Théâtre de l'Opéra in Paris was built in 1745, burned in 1763, reopened in 1770, and burned again in 1781.[8] French architect Adrien Louis Lusson remarked matter-of-factly in 1845 that theaters should be built in large open spaces lest the inevitable theater fire consume the entire neighborhood.[9] German underwriter August Fölsch reported 516 theaters destroyed by fire before 1877, including the Bowery theater in the United States, which burned down five times, and one Spanish theater that burned seven times in the same period. Once a theater fire went out of control, destruction was generally total within an hour.[10] The iron curtain Holland installed was a French invention, used by Jacques Germain Soufflot (1709–80) in a Lyons theater constructed in 1756, of which F.W.J. Hemmings tells us:

> It consisted of a large sheet of iron, cranked down from the roof and blocking the opening of the stage. It was intended for use only in emergencies; at the start it was kept well oiled so that it moved easily, but the manoeuvre proved cumbersome and was discontinued, and Soufflot's invention found no imitators in Paris until it was decided to incorporate a similar sheet metal safety curtain in the new Odéon, rebuilt in 1819. Since fires were more like to start on the other side of the footlights [that is, the stage side] this novelty would, it was hoped, reassure the spectators; but being enormously heavy, it could not be lowered with any speed. Later managers replaced it with a metal grid, which was lighter but of doubtful efficacy, and the Théâtre-Français reverted to a solid metal hanger as part of the additional precautions against fire intro-

duced in 1887. . . . Only with the growing use of asbestos in the twentieth century did a lighter curtain, and one moreover that unlike metal was not a conductor of heat, become a practical possibility.[11]

Frederick Carlisle, writing in 1809 about the burning of Holland's Theatre Royal as well as other theaters in London and elsewhere, remarked, "Owing to the miserable accidents which so lately occurred by the burning of the theatres of Covent Garden and Drury Lane, the opportunity is now afforded of noticing the many imperfections in the original building of our theatres, and of affording some security to him who goes to the play that in so doing he shall not walk into his tomb."[12]

The United States, while too young as a nation to have experienced the waves of bubonic plague that swept through Europe and Britain between the fourteenth and eighteenth centuries, wasted no time after achieving independence in contributing to the international plague of theater fires. Other kinds of building fires and conflagrations had already motivated some American communities to pass building laws of the kind that had been in place in parts of Europe since the twelfth century.[13] The use of thatching for roofs and of hollow logs as chimneys, "insulated" on their inner surfaces with dried mud, for example, was prohibited in Boston in 1631, which nonetheless had a major fire in 1653. New York (then New Amsterdam), apparently taking a somewhat belated lesson from its sister city to the north, banned thatch and wooden chimneys in 1656, even going so far as to order them removed from existing structures.[14]

Theater buildings, however, were not yet regulated in American cities in 1811. In that year Richmond, Virginia, lost sixty-eight of her citizens, including the governor of the state and a former member of Congress, when a chandelier on the stage was raised into the flies with one candle still burning. The destruction of this theater and 10 percent of its audience in what eyewitnesses report was less than twenty minutes is a paradigmatic example of a nineteenth-century theater fire, and several of its fatal elements were to persist in theater design until public outrage demanded the enforcement of strict building codes for theaters in the early twentieth century.

In the flies or rigging loft of a typical theater (the area immediately above the stage), a wooden or metal rack at the top of the stage house, called a grid or gridiron, supports rigging that raises and lowers the scenery drops (painted canvas scenes stretched on wooden battens), lighting devices, and any other equipment that is "flown," or suspended above the stage. The scene drops and flown lighting remain in the flies, behind the central portion of the proscenium arch and invisible from the auditorium side, until they are needed. When the scene in which they are used is complete, the stagehands "fly them out"—that is, raise them out of sight and fly in those needed for the next scene.[15] Theaters have been known to store hundreds of painted scenes in their flies in crowded rows very close together, plus the "flats" or "set-pieces" (stationary wood-and-fabric scenes), which are usually stored in the wings against the backstage wall.

Historically, stage scenery has been painted with the traditional mineral pigments known to artists, mixed with water, glue, or linseed oil. Of these pigments, several are significantly toxic and an inhalation hazard in a fire, among them realgar (red arsenic sulphide, As_2S_3), orpiment (yellow arsenic sulphide, As_2S_3), Naples yellow (lead [II]-antimonate, $Pb[SbO_3]_2$ or $Pb[SbO_4]_2$), emerald green (copper [II]-acetoarsenite, $Cu[CH_3COO]_2 \cdot 3\ CU[AsO_2]_2$), and red lead (lead [II, IV] oxide, Pb_3O_4).[16] Most theater fires begin on the stage; and when scene drops and flats ignite, these pigments form part of the combustion gases.

On the night of December 26, 1811, a Richmond, Virginia, stagehand, responding to an order from his superior after the first act of a pantomime, flew out the still-burning "lustre," or chandelier, that had lit the scene of a peasant cottage. The stage carpenter, belatedly realizing that he had ordered an open flame into the flies, seized the rigging and attempted to lower the lustre but mishandled it in such a way as to tilt the chandelier from the plane of the horizontal.[17] The candle flame promptly ignited one of the nearby scene drops, of which thirty-five, including some made of paper, hung side by side in the flies. All of them proceeded to catch fire one from the other until the flames reached the proscenium wall and the border curtain. At about this point the audience was apprised of its danger, first by the descent of burning material from the flies, and then by one of the actors, who announced the house afire. The stage carpenter and his helpers climbed into the loft and cut the burning scenery down, which had the effect of setting fire to the wooden boards of the stage, forcing those in the dressing rooms below to flee through the windows as there was then no means of reaching the backstage door.

No plans of the structure have survived, but close analysis of eyewitness accounts of the fire suggest that the Richmond Theater had at least three exits: the main entrance where white theatergoers presented their tickets, a "colored" entrance with its own fully enclosed stairway so that white patrons need not rub elbows with African Americans, and one or more backstage doors through which the players and the stagecrew entered and exited the premises. Set apart on its own hill above the town, the theater had no source of water for firefighting.[18]

Although the walls of the building were brick, the roof and the rafters supporting it were of unseasoned pitch pine, as were all of the interior finishes. This wealth of fuel had been traditionally accounted a virtue in theaters because wooden walls were thought to improve sound quality in the auditorium.[19] To prevent pine resin from dripping onto the fine clothes of the playgoers, old canvas fabric had been attached to the wood as a kind of makeshift drop ceiling. This fabric, of course, caught fire as soon as the flames spread past the proscenium arch, aided by the opening of the backstage door to permit the escape of the players and crew from the stage; and the rafters and the roof ignited from it almost at once, filling the auditorium and lobbies with heavy, choking, black smoke. Most of the fatalities from asphyxiation occurred at about this point in the fire's development. Typically in such narratives, eyewitnesses differ in many of their details, but there is consensus among them that "in less than five minutes the edifice, from the roof to the pit floor, stage &c., all was one entire sheet of fire."[20]

Because patrons descending from the boxes had to pass through the pit doorway, cross the floor of the pit, and go up the flight of stairs ominously known as the box staircase—in this case, appropriately enough, a pine box—to reach the box lobby, their way was almost entirely blocked by the audience escaping from the pit. Bodies were found piled in the box staircase after the fire had burned itself out. Unusually for a theater fire, many of the rich thus perished in the fire, while few of the middle and poorer classes did so, creating subject matter for sermons for decades thereafter on theater fires as divine vengeance on frivolous wealth. The African Americans, with their own exit, had so few casualties that abolitionist preachers soon seized on the fire as an example of God's judgment on slavery and slaveowners.[21] In the ninety-two years between the Richmond fire and the Iroquois Theatre's, newspaper editorials and public sentiment moved gradually away from regarding theater disasters as acts of God toward finding in them human agency or, more often, human negligence; and toward the end of the nineteenth century, fire-safety professionals, such as Captain Eyre Massey Shaw and Edwin O. Sachs in Britain and John Ripley Freeman in the United States were called in to investigate major fires.[22] Richard Allen Willis, noting that theater fires in the nineteenth century were rarely put out but "simply burned themselves out," adds that "[u]nder these circumstances, it is not surprising that audiences tended to panic. What is surprising is that audiences could be persuaded to enter the theatre at all." These accounts make it very clear why Justice Oliver Wendell Holmes asserted in 1919 that American freedom of speech does not extend to "falsely shouting 'Fire!' in a theater."[23]

Grim reports like that of the Richmond fire crossed the oceans in every direction for the rest of the century. On May 25, 1845, 1,670 Chinese playgoers died in Guangzhou (Canton) in the most deadly theater fire in recorded history. Nearly a year later a fire in the Royal Theatre in Quebec killed about two hundred persons. The Teatro degli Acquidotti in Livorno burned on June 7, 1857, killing at least forty-three people. About nine hundred more died in Shanghai in June 1871 in a theater built of bamboo and matting, and an estimated six hundred people died in Tientsin in May 1872. Mrs. Conway's Theater in Brooklyn went up in flames at about 11 P.M. one night in late 1876, with a loss of 283 lives. In a pattern already familiar to fire-safety professionals, a border curtain caught fire from a light batten, and there was no water available to extinguish it. When the actors and the crew opened the scene dock doors to make their escape, the fire exploded through the closed act drop into the audience, which had not been certain until that moment that the theater was on fire.[24] In Ahmednuggur, India, forty people perished in a local theater in May 1878, while to the east, five theaters burned in Osaka, Japan, during an eighteen-month period in 1878 and 1879. As we have seen, theater owners and architects had shown a sporadic interest in the safety of their audiences and players in the preceding century. Although cities that had suffered major-mortality theater fires, such as New York City, enacted stricter building laws in the 1870s, the global wave of theatrical destruction and carnage did not attract systematic international attention until two European theaters burned, with heavy loss of life, within seven months of each other in 1881.

The first of these disasters was the Municipal Theatre in the French city of Nice, set afire by defective gas fittings on the night of March 23, 1881, killing between 150 and 200 persons. The second was the true marker event in the regulatory history of public-assembly occupancies in Europe: the Ring Theatre fire in Vienna on December 8, 1881.[25] Like the theater in Nice, this building caught fire from gas fixtures, which were mishandled during the lighting process on stage at the beginning of the performance. The stage crew apparently panicked and made two fatal errors that doomed hundreds of members of the audience: they turned off the gas and therefore all the lights in the building, making the exits invisible, and they failed use the theater's principal defense against fire, an iron curtain similar to the one employed by Soufflot in Lyons more than a century before.

These curtains were extremely heavy, weighing seven to eight tons, dangerous to persons under them, and thus could not be made to descend safely without some human control, which in the case of the Ring Theatre's curtain could only be implemented from the rigging loft, where the fire was raging within seconds of the ignition of the hanging scenery. The exact death toll in the Ring Theatre fire is not known; estimates range from 450 to nearly 1,000, with a large proportion of the victims under the age of thirty. All accounts agree it was a catastrophe of sufficient proportions to at last impress on governments, the theatrical community, architects, and the playgoing public that theaters were scandalously fireprone and that something should be done about it.[26]

Some elements of what should be done were already known to theater architects by the seventh decade of the nineteenth century: theater owners and managers should not be permitted to pack more persons into the available space than could be evacuated quickly in the event of fire.[27] Distance of travel to exits should be minimized. Exit routes should not intersect in such a way as to cause evacuees to interfere with each other's progress. Exit passageways and stairs should be constructed of incombustible materials that do not produce smoke or toxic gases; doors should open outward.[28] Adequate firefighting and fire-reporting apparatus should be available throughout the theater building, and staff trained should be trained to use it. Roof ventilation should be provided above the stage so that smoke and gases travel upward and out of the building rather than across the footlights into the auditorium. Stage and house lighting should be separated so that a fire in the stage lighting does not require throwing the auditorium, vestibules, and exits into total darkness. And perhaps most important, it must be possible, quickly and reliably under adverse conditions, to make the stage and the auditorium into two entirely separate fire areas: first, so when a fire that breaks out where most theater fires start, on the stage, it does not spread to the auditorium as the Richmond fire did; and second, so that the evacuation of the actors and crew through exits behind the stage does not produce a draft that engulfs the audience in smoke, flames, and toxic gases. This second factor, essential to the proper functioning of all the other components of the safety system, was a lesson that had to be learned several more times, at a cost of hundreds of more lives, before adequate means were developed that completely and effectively closed the proscenium opening long enough for the audience to escape from a burning theater.

While engineers and architects struggled for nearly three more decades to learn from the failure modes of existing technologies for protecting theaters from fire, theatergoers, players, and stage crew continued to die. Roof vents, intended to be used in combination with iron safety curtains, were installed over stages in many theaters to vent the smoke and gases from stage fires upward to the outside rather than out into the auditorium.[29] In the aftermath of the Ring Theatre fire, experiments were performed on one-tenth-size models of the structure, reproducing various effects of the fire, including what they might have been had the iron curtain descended and there had been operational vents in the stage roof. Sachs reports some of the results of these tests:

> The first series of experiments disclosed how the gases on the stage expanded, blowing the proscenium curtain [the act drop] into the auditorium within seventeen seconds. It was next shown that, even before the auditorium was filled with smoke, the gaslights were extinguished by pressure of air from the stage; and further, that the highest pressure occurred within twenty seconds of the stage being well alight. The petroleum "emergency" lights were extinguished within twenty-nine seconds, whilst some "emergency" lights in which oil and candles were used were extinguished in thirty-one seconds. The air from the stage entered the gas-pipes, driving the gas back to such an extent that lights outside the auditorium were extinguished. . . . In another series of experiments, where two stage ventilators were provided, the results were very different, for, whilst the first ventilator was opened twelve seconds and the second ventilator twenty seconds after the outbreak, the reverse draft

> created caused the iron curtain to be drawn back on the stage, where it collapsed. . . . Every experiment was repeated three times.[30]

Clearly, the iron curtain invented by Soufflot and manufactured in Paris more than a century later was not as effective a fire barrier as safety experts had hoped. Holland and others, as we have seen, had already discovered the material's tendency to rust; others had noted their seven-to-eight-ton weight, their necessarily slow speed of descent, and the danger of their descending on persons underneath.[31] Gerhard summarized the problems of corrugated iron fire curtains in 1896 and went on to endorse the use of a new technology: curtains of woven asbestos mounted on or interwoven with a steel or brass wire framework and fitted with an asbestos-covered roll at the bottom to form a smoke seal. Charles John Hexamer also endorsed this type of curtain in 1892.[32] The technology for safety curtains was still, as Sachs expressed it, "in its infancy" in the late nineteenth century; and inventors, as usual, patented a number of devices of which few were workable or marketable.[33] Like Sachs, Gerhard condemned the wire-mesh curtains permitted by some European municipalities on the ground that they did not protect the audience from smoke, and contributed to panic because the fire could be clearly seen behind them.[34] Iron curtains, he said,

> should be accurately guided at both ends and should travel easily in proper vertical metal grooves placed on both sides of the proscenium-opening, and well fastened to the brick wall, otherwise they are liable to stick fast at the moment when wanted. Such iron sliding-curtains have not, in practice, proved themselves always reliable. It is important that the apparatus for raising and lowering the fire curtain should be on the stage, and not in the rigging-loft, as was the case in the Vienna Ring Theatre, because this point may be beyond reach soon after the outbreak of fire on the stage. Unless built very strong, iron curtains are liable to buckle and warp, and do not resist a strong pressure due to the expansion of the air by the heat in case of fire. They should fit tightly, to prevent the escape of smoke, and should be strong enough to sustain a pressure for at least ten minutes, to give the audience time to escape. If they are not protected from above by automatic sprinkling apparatus, or a series of perforated pipes connected with the supply-pipes or roof-tanks, so as to allow a sheet of water to run down along the curtain, they may become red-hot during a fierce fire and thus endanger the spectators.

A few pages later in his *Theatre Fires and Panics*, Gerhard reminds his readers that the time it takes for an iron curtain to descend can be very costly: "Incredible as it may sound, it is nevertheless true, that, on the average, only five minutes elapse between the appearance of fire in front of the proscenium-opening and the total extinction of human life by smoke and fire gases."[35] Note that the sprinkler system he mentions here, called a "water curtain," is only above the stage; an ordinary ceiling-installed sprinkler system of the type used at the time in mill construction would in a theater, as J. G. Buckley pointed out in 1888, be so high above the stage and audience as to be of very little service in a fire.[36]

At least one of the problems with iron curtains pointed out by Sachs and Gerhard had been amply demonstrated in May 1887, when the Opéra Comique in Paris burned with a loss of life estimated at 170.[37] Here again, the details are tragically familiar: a light batten ignited the scenery, and the resulting smoke and gases swept out into the auditorium. Although the Opéra was equipped with an iron curtain, the mechanism did not work, and the curtain could not be lowered.[38] In September of the same year, a fire at the Exeter Theatre in England killed 186 persons. Again, the fire began on the stage, quickly involved the hanging scene drops, and just as quickly asphyxiated and panicked the audience, there being no safety curtain of any kind or smoke vents above the stage.[39] In March 1888, a theater in Oporto, Portugal, went up in flames, killing another 170 people. The next month, the western theater community received word that the Korean National Theater in Kyoeng had burned with the loss of 650 lives. A fire at the opera house in Seattle, Washington, in June 1889 caused thirty deaths. A theater in Kamli in China set a new world record for fire mortality in April 1893, with an estimated 2,000 deaths. In the United States, a theater in Baltimore burned two days after Christmas 1896, with a death toll of twenty-two. A temporary theater in Paris, where a cinematograph was being operated in conjunction with a charity bazaar, had a terrible fire in 1897 in which an estimated 131 perished, the majority of them women, whose light summer clothing caught fire from falling fragments of the canvas used to the cover the ceilings of the newly built wooden structure. This disaster has the dubious distinction of being one of the first loss-of-life fires in a motion-picture theater, a venue that, as we shall see, was soon to acquire an evil reputation for such fires.[40]

Asbestos Takes the Stage

In the wake of these hideously repetitive tragedies, large cities in Britain, Europe, and the United States began to require safety measures in new theaters by the last decade of the nineteenth century; but it took one more major multiple-fatality theater fire before many building codes required the retrofit of existing theaters and began to mandate that unsafe theaters in the United States and Europe, regardless of their date of construction, be closed until code-compliant.[41] The exception was Rome: after the Ring Theatre fire, the Roman minister of public security ordered all major theaters in his city to be fitted with asbestos curtains without delay.[42] By the late 1890s, most large U.S. cities were requiring new theaters to have stage ventilators, fire alarms connected to the fire station by telegraph, and asbestos or metal fire curtains; the Metropolitan

Opera House in New York City, for example, had all of these features by 1897.[43] At the end of the first decade of the new century, metropolitan theaters in Britain and the United States could be shut down for failing to have any one of them, including a safety curtain with demonstrably effective performance in fire conditions that was at least equal to a heavy-duty, wire-reinforced asbestos curtain running in metal tracks on an incombustible proscenium wall, like those depicted in figs. 3.1 and 3.2. The fire that brought about this change took

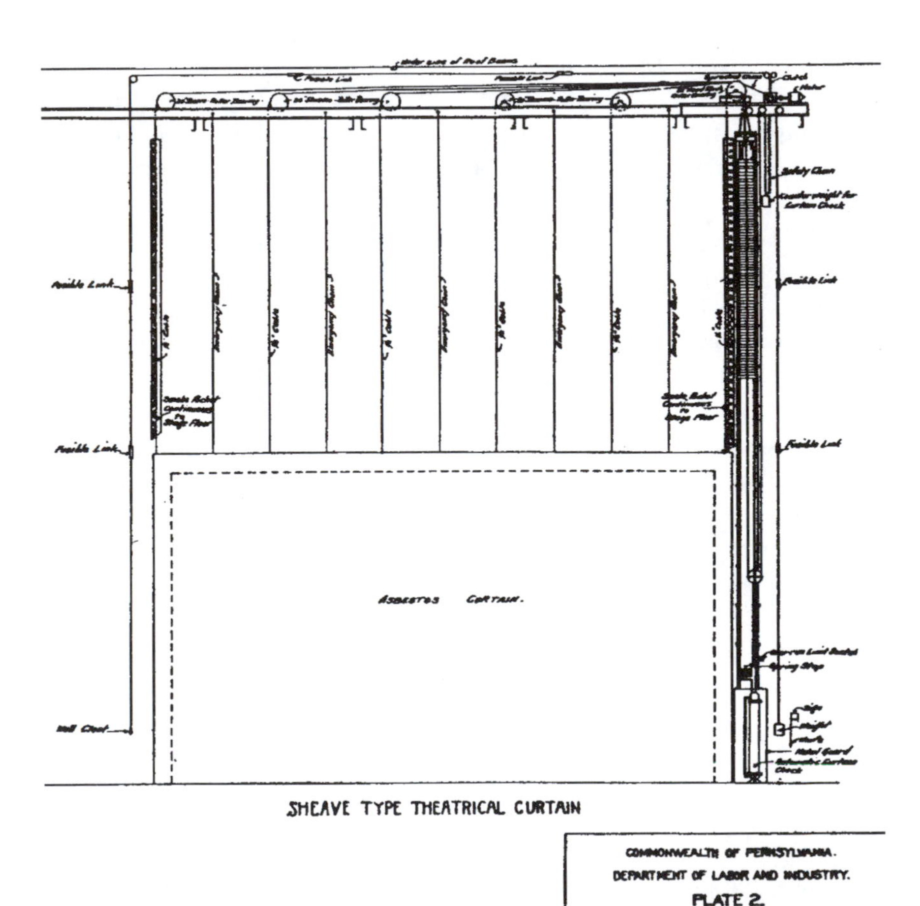

FIGURE 3.1 The sheave, or straight-drop, type of asbestos curtain was one of the approved types in American building codes after 1904. Unlike the roller type shown in fig. 3.2, this curtain was mounted on a rigid frame that required a proscenium wall somewhat higher than the curtain was long to be raised out of sight. This drawing was included in Pennsylvania's 1952 theater-safety code.

Pennsylvania Department of Labor and Industry, *Regulations for Protection from Fire and Panic* (Harrisburg: Pennsylvania Department of Labor and Industry, 1952), plate 2.

FIGURE 3.2 This steel-framed asbestos curtain by J. R. Clancy, Inc., of Syracuse, New York, was installed in the Purdue University Music Hall sometime before 1949. When raised, the sections rolled up onto a drum behind the proscenium wall.

Photo by J. R. Clancy, Inc. Reprinted by permission.

place in the Iroquois Theatre in Chicago, on December 30, 1903. This was the most deadly single-structure building fire in American history: 603 theatergoers died, twice as many as had perished in the Great Chicago Fire three decades earlier and more than all of the American fatalities in the Spanish-American War.[44] Forty percent of the total were children.

The Iroquois was a new theater, built, according to its architect, to the strictest standards of safety, including emergency exits that opened with complicated procedures in which the theater's staff had no training. Chicago already had a building code requiring an in-building fire telegraph, roof vents, fire-suppression apparatus, sprinklers over the stage, and an asbestos curtain. But in the rough-and-ready political climate of Gilded Age Chicago, the building managed to receive a certificate of occupancy and was opened before it was entirely completed. Some "minor" elements were missing in late December 1903: lighted signs indicating the exits, electrical separation of house and stage lighting, firefighting equipment and training, and operational roof vents. These last had been completed but were still nailed shut from the outside. The theater's owner later claimed to have complied with the regulation regarding the asbestos curtain; but in the absence of specifications in the code as to the quality and type of curtain, he had economized and bought what Sachs later called an "unsuitable loose cloth drop-curtain, or so-called reefed act-drop type, temporarily mounted on wooden sheaves and worked by ordinary ropes. There is no record of its regular use or trial."[45] Later, the remains of this curtain were found to consist mostly of vegetable fiber.[46] There were no satisfactory explanations for the lack of a fire-alarm telegraph box or the fact that the roof vents were still closed six weeks after the theater's first performance. The only ventilation for smoke and gases was in the back of the auditorium, directly over the heads of the audience in the gallery. The response to a later inquiry about sprinklers, which were required by the code, was that no Chicago theater of the time had sprinklers. It was later discovered that only one city theater had a true asbestos curtain, the others having faked compliance using materials as obviously combustible as canvas and burlap.

The fire started at about 3:20 P.M., during a matinee performance of *Mr. Bluebeard*, when one of the arc lamps ignited a border curtain, which, as in all the previous cases I have described, set fire to the 3,000 square yards of canvas scene drops stretched on 8,000 board feet of white pine in the rigging loft, in which hung eleven miles of manila rope. The stage fireman climbed up to the border in the early stages of the fire and used the only firefighting equipment on hand: two tubes of powder fire extinguisher, which had no discernible effect. One of the flymen was asked to lower the border so that the fire could be stamped out on the stage, but instead he dropped the "asbestos" curtain, which then snagged up on one side, like a Roman window shade improperly let down, on a reflector that had temporarily been extended past the proscenium arch in the previous scene. One of the players, Eddie Foy, went out under the partially closed curtain and urged the audience not to panic.[47] While he was reassuring them, pieces of flaming scenery began to fall around him, and prudent souls made for the exits. This was not easy because the theater's emergency exits had been draped with fabric for aesthetic reasons and there were no exit signs. The stairs from the gallery were partially closed off with an accordion gate to prevent impecunious playgoers from shifting to the more expensive seats below. Most of the doors opened inward.

Those who managed to find the artistically camouflaged exit doors discovered that their European-style bascule locks opened by some mysterious principle not obvious to persons fleeing for their lives with small children in their arms. Many bodies were later found piled around these doors. Another artistic embellishment, a set of mirrored false doors, also proved to be a deathtrap, although the worst of the Iroquois's architectural deceptions was a hallway that apparently led to an exit but in fact dead-ended at a blank wall. A large pile of bodies was later found in this corridor. Some of the exit doors, including two at the main entrance, were locked. Stairs from the gallery and the first balcony converged at a right angle so that the streams of persons fleeing each level collided at the intersection.

Meanwhile, in the auditorium, those who had remained in their seats were about to take their last breath. As the actors and the crew escaped from the building, the rush of air from the scene dock doors backstage sent a rolling wall of flame out under the half-lowered "asbestos" curtain, blowing the fabric straight into the auditorium, where it shattered, while flames and gases streamed directly toward the vents in the back of the gallery. Seventy percent of the mortality in this fire occurred in the gallery, where the recovered bodies were apparently immolated or suffocated without having left their seats, in many cases with one arm held up before their faces. When the burning scene drops and rigging crashed to the floor of the stage, they struck the electrical switchboard, casting the entire building into darkness, increasing the difficulty for survivors still trying to get out of the building.

The alarm box for the fire was a full block away, so the fire burned for more than thirteen minutes before firefighters were summoned. Once called, they arrived by 3:40 P.M., just as the fire was burning itself out. What remained of the blaze was quickly extinguished, but 582 lives had ended in a quarter of an hour. The remainder of the butcher's bill was accounted for by later deaths from injuries and burns. The Chicago coroner's office did not attempt to determine cause of death for most of the fire victims or even to keep track of all the bodies, some of which were claimed on the sidewalk outside the theater. The building had been advertised as "Absolutely Fireproof," and so it was: damage to the building was so minimal that it was able to reopen less than nine months later. But in Chicago there was no celebration of New Year 1904, which arrived a day and half after the disaster.

Like all such catastrophes, the Iroquois fire galvanized the international fire-safety community. Chicago's mayor closed all the city's theaters pending safety inspections; in New York City officials went from theater to theater with cans of gasoline and matches to test the safety curtains, closing theaters that were noncompliant.[48] The incidence of heart failure among stage managers whose curtains were tested in this way is not known. A few years later, the New York City Fire Department built a scale model of a state-of-the-art fire-safe theater, complete with operational asbestos fire curtain, for display at the Panama-Pacific exhibition.[49] One of Chicago's theaters installed two-ton safety curtains of a new type: steel backed with asbestos millboard, the folding shutter curtain having been invented late in the previous century, primarily as a security measure for stores with plate-glass windows, an application for which it is still the dominant technology. By 1909, thirty-three Chicago theaters were said to have steel and asbestos fire curtains.[50]

Staunton, Virginia, closed its opera house pending a bond issue to renovate the second-floor facility.[51] Good asbestos safety curtains for theaters were expensive, costing $175 to $600 each, plus installation, in 1906, depending on the size of the theater and its proscenium opening.[52] Keasbey & Mattison Company, one manufacturer of these curtains, wrote in 1924:

> There is no application of Asbestos Cloth where the requirements are more rigid than in the manufacture of Asbestos Theater Curtains, which must be fire-resisting in the highest degree. The cloth must be of close weave, it must have strength, and it must have weight.
>
> The use of plain Asbestos cloth is no longer regarded as good practice. A metallic cloth weighing five pounds to the square yard most effectively meets the requirements. In some States a three-pound cloth is acceptable and others permit the use of two and one-half pound cloth.[53]

As this advertisement suggests, after the Iroquois fire, American municipalities began to be far more specific in their building codes about the required quality, method of installation, and materials of theater safety curtains. Section 81 of the building code of Dayton, Ohio, enacted in 1916, described the required "Proscenium Curtain" in considerable detail:

> a. Curtain—The proscenium opening in theaters containing movable scenery shall be provided with a fireproof rigid curtain, or a curtain of asbestos or other fireproof material, approved by the Chief Inspector. If the proscenium curtain is of asbestos, the same shall be of pure asbestos fiber, shall weigh not less than three (3) pounds per square yard and shall be reinforced with wire or wire spun in the asbestos. At the top and bottom of the curtain shall be placed a wrought iron or steel pipe not less than two (2) inches in diameter, securely fastened to the curtain and covered over with like material as the curtain itself. No oil paint shall be used on any asbestos curtain. Curtain shall overlap the brick or concrete proscenium wall at each side not less than twelve (12) inches and at the top not less than two (2) feet.
>
> b. Channels and Supports—The curtain shall slide vertically at each side within iron grooves or channels to a depth of not less than twelve (12) inches, said grooves or channels to be made of iron or steel not less than one-quarter (1/4) inch in thickness, be securely bolted to the brick or concrete wall, and extend to a height of not less than one (1) foot above the top of the curtain when raised to its full limit. The curtain shall be suspended or hung by steel cables passing over the ball bearing iron or steel sheaves supported by wrought iron brackets of sufficient strength and well braced, and brackets shall be securely attached to proscenium wall by through bolts with nuts and washers on the other side of the wall.
>
> c. Checks—The excess weight of the curtain shall be overcome by check ropes of cotton or hemp, extending to the floor on both sides of

the stage so that the cutting or burning of which will release the curtain and the same will then descend at its normal rate of speed.[54]

In Dayton and some other municipalities, the older type of metal fire curtains were still permitted, and indeed this older technology has persisted in archaeomorphic theatrical argot: "iron going out" still means that the safety curtain is being drawn or raised.[55] The burden of showing that these alternatives were equal to the protection of asbestos was, however, on the property owner until the U.S. Bureau of Standards tested both asbestos and steel fire curtains in 1924 and 1933 and gave a clean bill of fire-safety health to both materials, but not to iron.[56] Steel curtains were considerably more expensive than asbestos and slower to raise and lower. Because of their great weight, they required power-assisted lifting apparatus and counterweighting. Large theaters installed safety curtains of both materials, a steel curtain facing the audience with a thick, framed backing of asbestos millboard to prevent the steel from heating up and failing in a stage fire. Sanborn fire-insurance maps of American communities noted on theater building icons whether or not the structure included an asbestos curtain.[57] That audiences appreciated the importance of these measures after the Iroquois fire is evident from a cartoon published in 1904, depicting the Russo-Japanese War as a stage play watched by other nations from what they clearly hope is a safe distance; Uncle Sam is shown asking whether or not the asbestos curtain is in working order (see fig. 3.3).

Most theaters, however, including small local and institutional theaters, such as those in schools, opted for one of several types of asbestos curtain, classified by weight and "action"—that is, how they were opened and closed. Once the safety curtain became standard in theaters, it became a tradition to make it aesthetically pleasing and sometimes also to use it as a kind of advertising billboard for local businesses. Most of the pre-1950 curtains were prominently labeled "asbestos" both to reassure audiences of their security and because some jurisdictions required this measure to prevent panics, which experience had shown could be almost as deadly as fires.[58] Some of these curtains survive in place in theaters, and a few others are in museum collections.[59] "The asbestos," like its predecessor, the iron curtain, also became a metonym for safety curtains generally; and the archaeomorphic technical term is still used today by stage folk in the era of fiberglass curtains.[60] Safety curtains had an impact on the artistic trajectory of theater, as Simon Tidworth notes: "the separation of stage and auditorium by a fire-proof safety curtain . . . had the effect of preserving the proscenium arch long after many architects and directors wished to discard it."[61]

Most important, the slaughter of audiences in the live theater ended in the United States with the Iroquois. Stage fires continued to occur, including the newer type of fire caused by electrical faults, but they no longer killed large numbers of playgoers. In the hundred years before the rise of the asbestos

ALL—"IS THE ASBESTOS CURTAIN IN WORKING ORDER AND ARE THE EXITS IN PERFECT CONDITION?"

Drawn by F. I. Leipziger, of the Detroit Evening News.

The fierce battle scene is fairly on at the International Theater, with the Mikado, Uncle Sam and the great nations of Europe in the front seats watching the performance. The horrors of the Iroquois Theater fire are still in the public mind.

FIGURE 3.3 This cartoon of the Russo-Japanese War by F. I. Leipziger of the *Detroit Evening News* recalls the then-recent disaster at the Iroquois Theatre in Chicago. The original caption reads, "All—'Is the asbestos curtain in working order and are the exits in perfect condition?'"

"Russo-Japanese War Cartoons (1904–05)," *http://faculty.Kirkwood.edu/ryost/russonjapanese.htm.*

curtain, about a hundred persons worldwide, on average, died every year in theaters. In the century since then, despite the fourfold increase in world population, the average global theater death rate has dropped to about twenty-five persons per year, with no triple-digit theater fatalities in the United States, Britain, or any western European nation since 1908.[62] Asbestos curtains were required in many twentieth-century U.S. municipal codes and in some state codes, such as Pennsylvania's.[63] The asbestos safety curtain was a highly successful component of the system of theater safety devised by late nineteenth- and early twentieth-century fire-safety professionals.

Predictably, one of the responses to the Iroquois Theatre fire was a wave of inventions of supposed improvements to safety-curtain technology, some of which, such as hydraulic lifting mechanisms, actually worked as described and became part of later theater-safety technology. Others, like the mica fire shield invented in 1906, sank into historical oblivion with scarcely a trace.[64] Despite the competition from steel, asbestos emerged in the early twentieth century as the dominant material for theater safety curtains and was to play a similar role when theaters began to recognize the dangers of a new entertainment technology: the motion-picture film.

The last major theater fire in the United States was the Rhoads Opera House in Boyertown, Pennsylvania, where a lantern-slide projector was being used to project scenes of *The Scottish Reformation* in January 1908. One hundred seventy persons in an audience of a little more than three hundred lost their lives after one of the fuel tanks for the calcium light in the projector exploded. Somehow in the next few minutes the kerosene tanks for the stage footlights also ignited, engulfing the second-floor auditorium in flames. The stage exits and stairs allowed the cast to evacuate, but the row of burning footlights prevented the audience from using this escape route. A few remembered the windows and the fire escapes under them, but most headed for the main door, which opened inward, with predictable results.[65]

The danger of operating projectors in the same fire area with an audience was clearly shown in this catastrophe; but with the introduction of nitrocellulose motion-picture film, a new and more deadly element was added. In his 1932 study of Canadian provincial regulations for motion-picture film, Richard Ruedy summed up the hazards of working with it:

> The dangerous features in the combustion of nitro-cellulose are first, the energy set free when the material is heated and decomposes and the rapid rate of combustion, which is about ten times as fast as wood or paper; second, the high oxygen content which allows nitro-cellulose to burn without flame in the absence of atmospheric oxygen; and third, the poisonous gases and the explosive gases given off when the combustion is incomplete.[66]

Running this material in short jerks past a high-intensity arc lamp was, of course, a very risky business; and if the film broke, a fire could break out that would quickly engulf the entire building if it were not contained within a fire-resistive projection booth. Breathing the fumes of the burning film was often fatal, and large quantities of rolled film can explode violently enough to cause third-degree burns at a distance of half a mile. Invented in 1889, nitrocellulose film was the subject of safety and licensing legislation in Britain within twenty years, after a film warehouse in Pittsburgh burned in a spectacular blaze that killed ten people and injured another twenty.[67] The Cinematograph Act of 1909 included a provision that all public motion-picture projector booths in the United Kingdom be fire resistive. Section 19 of the act specified, "The enclosure shall be built of brick, tile or plaster blocks, plastered on both sides, or of concrete, or of a rigid metal frame, properly braced, and sheathed and roofed with sheet iron of not less than No. 20 U.S. gauge metal, or with ¼-inch hard asbestos board, securely riveted or bolted to the frame, or 2 inches of solid metal lath and cement or gypsum plaster."[68]

By 1910, the National Board of Fire Underwriters had developed the initial procedures and standards for handling nitrocellulose film that were later to become NFPA 40, and the National Fire Protection Association had begun a forty-year campaign to have all films distributed on safety stock.[69] The state of Michigan had a moving-picture safety and licensing ordinance in place by 1913, which required asbestos and metal projection booths. Theaters that could not meet the safety standards were closed by the state inspectors.[70] In his 1916 *Theatres and Motion Picture Houses*, architect Arthur Meloy describes the options available for projection booths (brick, concrete, terra-cotta, and so on) and points out that asbestos booths could be ordered prefabricated in sections, suggesting that there was a significant market for them. He tells his readers, "The floor of the projection room must also be fireproof. If the booth rests on a wooden platform, it must be covered with asbestos lumber 3/8 in. thick."[71] Projection booths of this code-compliant type are depicted in figs. 3.4 and 3.5; the latter will be familiar to moviegoers as the probable model for the post-fire projection booth in the Italian film *Cinema Paradiso*. Reports of fires after these

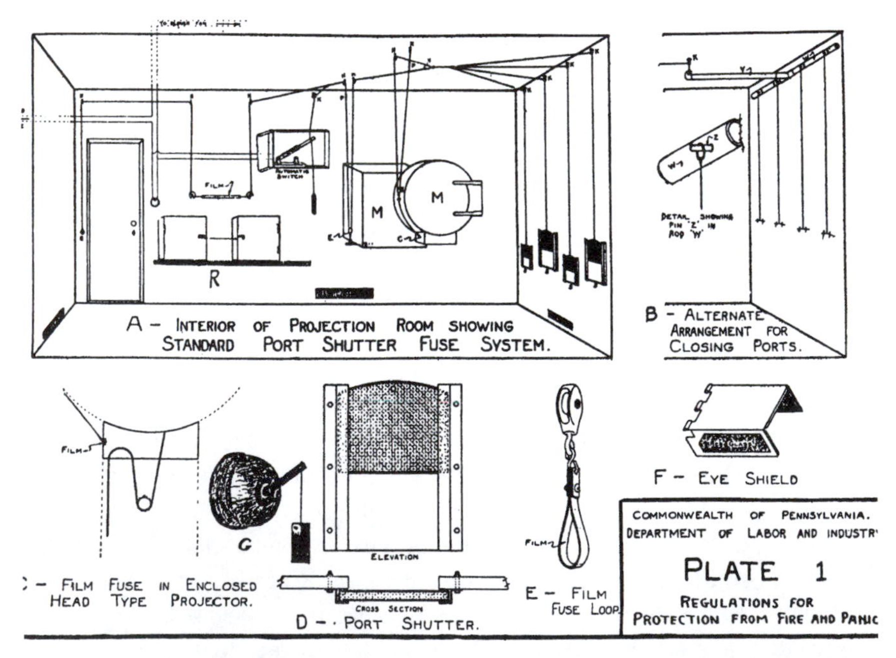

FIGURE 3.4 This 1952 drawing by Pennsylvania's Department of Labor and Industry shows the proper layout of a fire-resistive motion-picture projection booth. To comply with the state code for motion-picture theaters, all interior surfaces of these booths had to be asbestos or metal.

Pennsylvania Department of Labor and Industry, *Regulations for Protection from Fire and Panic* (Harrisburg: Pennsylvania Department of Labor and Industry, 1952), plate 1.

FIGURE 3.5 The light-colored walls in this projection booth, circa 1910, are almost certainly made of asbestos. Portable projection booths were built almost exclusively of asbestos board on metal frames at this period.

F. H. Richardson, *Motion Picture Handbook*, 3d ed. (New York: Moving Picture World, 1916).

code reforms in the *NFPA Quarterly* showed asbestos to be an effective material for preventing and containing projection-room fires; and it was recommended in several applications, including as electrical insulation, in F. H. Richardson's *Motion Picture Handbook* of 1910.[72]

If anyone still had any doubts of the danger to audiences from projecting nitrocellulose film within the auditorium fire area, two multiple-fatality fires in the 1920s brought the lesson home. The first was a hideous episode in the Irish village of Dromcollogher, County Limerick, on September 5, 1926, in a "theater" consisting of the second floor of a grain store with stone walls and wooden floors and roof, from which the only exit was a wooden ladder on an outside wall. One hundred and fifty persons were crowded into the theater, with the projector between the audience and the exit door. When the film caught fire in the projector, flaming pieces ignited other films lying on a nearby table, and within seconds, the entire building was alight. Forty-eight bodies were pulled from the ruins.[73]

The second was the 1929 Glen Cinema fire in Paisley, Scotland, in which seventy children lost their lives. In the same year, 4,200 pounds of smoldering nitrocellulose X rays stored at the Cleveland Clinic sent toxic smoke into the hospital ventilation system, killing 129 patients and staff members.[74] All three episodes set off new international cycles of both model and enacted building code revisions. As Ruedy documents, by 1932 some Canadian provincial governments required or recommended asbestos lining of projection booths, with two providing detailed instructions for building the booth to code, and most large U.S. cities had building codes requiring projection-booth enclosures of either metal or asbestos or both.[75] Model codes nearly always specified asbestos for projection booths.[76] Some states, including Pennsylvania, also passed laws requiring fire-resistive projection booths as early as 1929 in order to regulate theaters and auditoriums outside municipal fire limits.[77]

In 1937, reinforcing the lessons of the preceding decade, the United States–Fox Film Storage facility in Little Ferry, New Jersey, burned to the ground. So combustible was nitrocellulose film that it became a metaphor for something that would burn in an instant, as in the popular expression, attributed to Billy Sunday, "to have as little chance . . . as a celluloid cat chasing an asbestos dog in hell."[78] In the same year, the *Uniform Building Code* of the Pacific Coast Building Officials Association devoted eight pages of small type to specifications for asbestos curtains and other fire-safety measures in theaters.[79] By the time cellulose acetate safety film became standard in theaters, the danger of motion-picture fires had been greatly reduced by these containment measures.

Fire-Resistive Materials in Building Codes, 1899–1941

The international movement to make theaters safe converged with ongoing work on other aspects of fire risk in the built environment at the end of the nineteenth century, including those of schools, ships, and other public-assembly occupancies such as hotels and workplaces as well as that of urban conflagrations, which had already been a major focus of municipal regulation for at least seven centuries. Issues associated with combustible roofing were prominent in European codes, with most banning thatch and wooden shingles entirely from built-up areas.[80] France and some other European nations had municipal codes that made homeowners responsible for the value of neighboring structures if burning brands from their wooden-shingle roofs ignited them, up to the value of an entire city block.[81] As a motivation for property owners to roof their buildings with fire-resistive slate or tile, and later asbestos, these laws were highly successful and were in fact adopted in the early twentieth century by some American municipalities.[82]

Wooden roofs were a major issue in the United States and Canada. Both nations, being heavily forested as Europe and Britain no longer were, had (and indeed still have) strong timber lobbies and active public-relations organizations that have historically opposed any restrictions, local, state, or federal, on the use of wood in construction.[83] The National Fire Protection Association reported in 1935 that wooden roofing shingles were a contributing factor in fifty-five of one hundred American urban conflagrations between 1900 and 1934 and that in one city and one year—Minneapolis in 1934—sparks or brands on wooden roofs caused 785 fires.[84]

The wave of U.S. urban conflagrations between 1870 and 1911, described in chapter 1, was a strong impetus to systematization and codification of existing knowledge about preventing and suppressing structural fires. North America's wooden cities burned at an alarming rate at the turn of the nineteenth and twentieth centuries, not only because urbanization had made these cities larger and more congested but because new materials and processes, such as electricity and combustion engines, added threats to the fire-risk environment that were still unfamiliar to fire-safety professionals.[85] Even in Britain, which had a much lower rate of urban fires than did North America, the British Fire Prevention Committee embarked on a systematic program of testing construction and other materials for fire resistance, completing tests of asbestos floors under fire conditions in 1898, ceilings in 1902, partitions and walls in 1905, fire doors in 1906 and 1909–10, cloth extinguishers in 1908, and roofing in 1913.[86]

Some aspects of fire risk in urban environments were fairly well understood, particularly those involving the use of wood for heating, cooking, and building: wooden roof coverings, chimneys, and stoves that burned solid fuel. Asbestos appeared early in the twentieth century in municipal and other codes in the United States, Canada, and Britain that regulated these uses of combustible materials in the fire districts of cities and towns, usually specified in combination with metal. Metal alone in, for example, a fire-resistive plate between a woodstove and a wooden floor, could heat to a temperature that would ignite the wood beneath, so the code construction specification was typically for a sheet of asbestos in some form, usually millboard, under the sheet metal and in contact with the floor, to reduce thermal conductivity.[87] Pipe insulation, including wrapped asbestos lagging, became a common requirement in building codes, especially after the floor of the Lakeview Elementary School in Collinwood, Ohio, now a suburb of Cleveland, caught fire from a furnace or one of its pipes and burned, killing 176 children in 1908.[88] Fire-safety professionals warned builders to use only real asbestos, not cheaper combustible look-alikes.[89] As *Textile World* expressed it in 1914:

> The marked increase in disastrous fires is directing more attention every day to the need of fireproof building materials that can be relied upon. The failure of so many so-called fireproof materials when subjected to the intense heat of conflagrations suggests the need of more careful judgment in the choice of these materials as well as a more stringent interpretation of fireproof building regulations.

> The demand for building materials that would not be affected by fire has encouraged manufacturers to experiment with all kinds of materials, but the only one that has thus far successfully withstood all tests is asbestos.[90]

Electricity, while generally considered safer than gas, oil, or candle lighting because it did not require an open flame, began to be recognized as a significant ignition hazard within a few years of its introduction into urban environments.[91] As insurance underwriters had little or no experience of how the use of electricity altered the conditions of actuarial risk or what materials and constructions would reduce this risk, the Underwriters' Laboratories (UL) was created in 1894 to test and certify materials, assemblies, and constructions according to published standards of safety. Harry Chase Brearly, the metallurgist who invented stainless steel, wrote in 1923:

> No sooner have we seized upon some new facility than we are likely to learn that nature may exact a serious price for its use. One evidence of this is found in fire losses which, in the United States, *increased more than one thousand per cent.* between 1865 and 1922, while the population increased but two hundred per cent. A study of fire causes shows that a large part of that loss can be traced to comparatively new devices and processes. [emphasis in the original][92]

Asbestos products of various kinds were rigorously tested; and during the half-century after UL's founding, a large number were approved for thermal and electrical insulation, building construction, and other applications. These UL-approved assemblies, such as one for brick chimneys with asbestos firestopping, were recommended and sometimes illustrated in model safety codes, such as those of the NFPA, whose members included fire-safety officials at many levels of government in the United States and Canada, and the Pacific Coast Building Officials Conference, an organization for commercial and industrial property owners and managers.[93]

The NFPA also tested and approved individual products, as in the case of asbestos building lumber in 1909; and contributors to its publications made recommendations later incorporated into codes, such as Edward Keith's recommendation in 1907 for lining drying rooms with asbestos millboard.[94] Organizations such as the U.S. Bureau of Standards, whose work I have already mentioned in connection with theater curtains, and the American Society for Testing Materials (ASTM), assisted in this process with testing programs of their own and by establishing standards and test procedures that supplemented and complemented those of fire-safety and engineering organizations (see appendix B).[95]

Safety standards and model codes were written by committees within professional organizations such as the National Conference of Building Officials and the NFPA and were subject to intense peer review and criticism during the process of arriving at consensus. The proceedings of, for example, the NFPA annual meetings testify to the vigor, and sometimes the acrimony, of debate over the establishment of safety standards. Only when consensus was reached would a provision become part of the model code or be accepted as a named and numbered standard. Consensus is still the method by which these codes and standards are updated and revised.

Most municipalities adapted one or more of the model codes to local conditions and enacted building laws that contained the provisions and recommendations that had passed through this, as it were, trial by organizational fire. As asbestos gained the confidence of the existing professional communities of fire-safety concern, it was recommended more and more widely in model codes and required in building construction by some insurance codes, such as Factory Mutual's, as a prerequisite to insuring the building.

After 1900, it became increasingly difficult to obtain a building permit in the United States for any building within the fire limits of a municipality without including at least some asbestos in the construction, wiring, or interior finish. Even outside municipal fire limits, state building codes drawn from the same sources applied to many structures. Codes used by insurance companies such the National Board of Fire Underwriters made it difficult to insure, and therefore to finance and build, commercial, residential, or industrial buildings unless provisions were made for proper insulation of fire hazards. Banks would not and will not make a construction, mortgage, or improvement loan on an uninsured structure.

In some applications, as in theater curtains and under stoves that burn solid fuel, asbestos was explicitly required by some codes; there was no acceptable equivalent. In others, such as roofing, ceiling tiles, pipe and boiler insulation, marine bulkheads, and some kinds of electrical insulation, it was one of several approved options. Tile roofs, for example, were thought to be as good as asbestos for fire resistance but could not be laid on buildings not designed to carry their weight, which was much greater than that of wood, asphalt, or asbestos. Asbestos had been an approved roofing material for mill construction in the late nineteenth century and, after testing in both the United States and Britain, was incorporated into building codes as an underwriter-approved type in the twentieth century.[96] By 1929, H. C. Jones, Jr., could write that, in mill villages, "[t]he hazard from the wooden shingle is being gradually eliminated in most villages by replacements with asphalt composition or asbestos shingles when reshingling is done."[97] According to tests by Underwriters' Laboratories and the National Bureau of Standards in the first half of the twentieth century, asbestos was the only cost-effective insulating material with a flame-spread ratio of zero. It remains the standard for zero flame spread defined in ASTM E84, the groundwork for which was laid in the invention of the Steiner tunnel test at Underwriters' Laboratories in 1911 (see fig. 3.6).[98]

Other dangers were under investigation by the international fire-safety community in the same period. New sources of ignition such as automobiles, aircraft, steamships, locomotives with their spark-spitting stacks, and fireprone industrial processes such as arc welding were receiving attention in professional journals.[99] A disastrous fire in the Paris metro in 1903, in which eighty-four persons lost their lives, was traced to the sparking of ferrous-metal brake linings in the tunnel and station areas, prompting both the adoption of asbestos brake linings for urban underground transit and automobiles as well as the elimination of combustible materials, including wood, from the exteriors of subway cars in most cities. Truss plates, third-rail covers, and car flooring in light-rail transportation were also made of asbestos.[100]

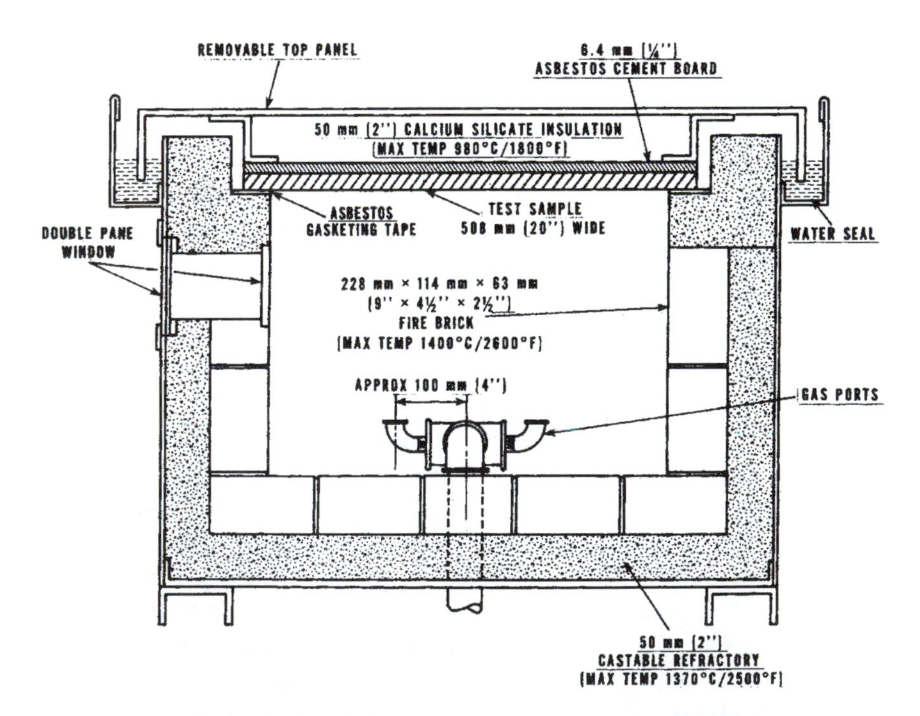

FIGURE 3.6 Drawing of a Steiner tunnel. Flame spread is measured by how quickly flames reach the ends of the tunnel, compared with the zero point of the asbestos-cement panel against which the tunnel was calibrated.

William J. Parker, *An Investigation of the Fire Environment in the ASTM E84 Tunnel Test* (Washington, D.C.: National Bureau of Standards, Center for Building Technology, 1977).

While children did not fall into hearth fires as often as they had in the nineteenth century, twentieth-century urchins had greatly expanded access to matches and later to cigarette lighters, introduced wholesale into the environment by that walking fire hazard, the smoker of tobacco.[101] Cigarette smoking was regarded by fire-safety professionals as particularly pernicious because, unlike pipes or cigars, cigarettes will burn unattended and thus will more readily set fire to furnishings, dry vegetation, and other combustibles.[102] The deadliness of smoking cigarettes while handling nitrocellulose film was observed on more than one occasion.

In cases of this kind, it was not only the sources of ignition that concerned fire safety professionals; the fuels, too, were new, some of them more combustible than wood and poorly understood as to their volatility and effects under oxidizing conditions. Fires in some of the new fuels, such as gasoline and motion-picture film, could not be readily extinguished with water. For all of these new ills as well as for the old, the best cure was, predictably, prevention. Building codes and manufacturing standards were devised to prevent fires from both new and old sources, and in many applications asbestos was endorsed by professionals as the best-available material for achieving this result.

The 1933 city of Cincinnati building code, which was in most respects typical of interwar-period U.S. municipal building codes, had the usual theater safety curtain requirement for "a rigid curtain of asbestos or other *approved* fire resistive material" [emphasis in the original], in which the "curtains shall be constructed of pure asbestos fiber interwoven and reinforced with wire strands, and shall weigh not less than 3 pounds per square yard," plus electrical insulation made of asbestos in motion-picture projection equipment. The mineral was one of three roofing materials approved for the fire district and was required in heating-appliance insulation as well: "Where protection is *required*, the same shall be by asbestos board at least 1/4 inch thick, plaster board 3/8 inch thick, or bright sheet tin covered with 2 thicknesses of 10 lb. asbestos sheet or an *approved* equal" [emphasis in the original]. As always, the burden of proof of equivalence was on the property owner. Wooden construction around chimneys "shall be insulated from the masonry by asbestos paper, at least 1/8 inch thick, and metal wall plugs or *approved* incombustible material to form a firestop" (emphasis in the original). In laundries and other establishments that used heat for drying, "No dry room constructed of incombustible material shall be placed within 18 inches of any combustible material unless such combustible material is shielded by at least 1/4 inch asbestos building lumber or hard asbestos board." When acids or acid solutions were carried from buildings in pipes, "All joints in acid proof composition metal pipe and fittings shall be made with asbestos packing and poured lead not less than 1 inch deep or standard screwed joints. All joints in acid proof chemical stoneware or glass pipe and fittings shall be made with asbestos packing and poured bituminous compound." Many municipalities had similar provisions, which were of course drawn mainly from model codes.[103] Drawings like those in fig. 3.7 were included in many municipal and model codes to ensure that instructions for constructions and materials were clear and unambiguous.

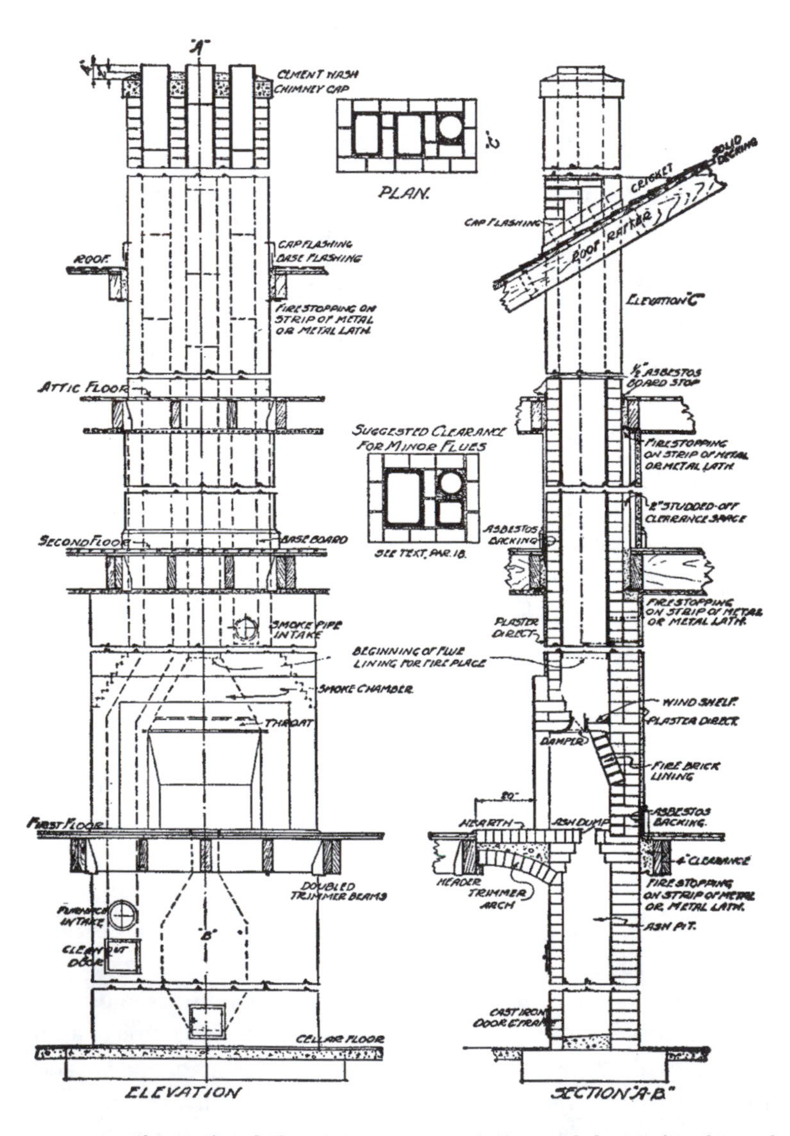

FIGURE 3.7 The National Fire Protection Association and the National Board of Fire Underwriters published many diagrams of standard constructions that included asbestos, as this one does in the chimney firestop at upper right. This drawing was reproduced often in municipal and state code manuals.

Reprinted with permission from the *National Fire Protection Handbook*, 10th ed., Copyright © 1948, National Fire Protection Association, Quincy, MA.

As more knowledge was gained about fire prevention and suppression in the twentieth century, the body of knowledge represented by building codes and fire safety handbooks became visibly larger, as did the volumes in which they were published. Not only were more subjects added, but existing sections were expanded as more was learned about the prevention and suppression of fire. By the time the eighth edition of the *NFPA Handbook of Fire Protection* was published in 1935, consensus standards had created an impressively complex and effective system of fire safety that included fire-resistive materials for construction and interior finishes, adequate exits, management and operation of trained professional firefighting units, telecommunication of alarms, and various methods of active fire suppression within structures.[104] As with all such best-laid plans, the chief difficulty was and is to implement and enforce them on a day-to-day basis.

Sprinkler systems, first used in the nineteenth century to protect property in factory and mill fires and later to help safeguard life in venues such as nursing homes, were part of this system; as a technology they developed contemporaneously with that of asbestos materials. Fire-resistive materials are intended to deny fuel to a prospective fire; sprinklers are intended to suppress a fire that has already found material to consume. Their principal drawback has always been their vulnerability to human error. From 1911 through the 1940s, the *NFPA Quarterly* routinely ran a column called "Unsatisfactory Sprinkler Fires"—that is, fires in which the sprinkler system failed to suppress the fire.[105] By their nature sensitive to factors such as the heat that activates them, sprinkler heads could be and were rendered inoperable for all sorts of reasons: freezing of pipes in the winter, wasps' nests, sabotage, inappropriate location in the building with respect to possible sources of ignition, incompetent installation, and even water shutoff due to nonpayment of the water bill. In 1923, Underwriters' Laboratories was maintaining a museum of sprinkler heads that had failed to open under fire conditions, which already covered most of a wall.[106] The passive components of the fire-safety system, such as the use of fire-resistive materials in construction, are not as readily derailed as active fire suppression by the historically intransigent human tendency to discount the possibility of imminent disaster. Sprinkler systems were and are necessary for fire safety, but they are by no means sufficient. As fire scientist Kimberly D. Rohr recently observed:

> Even a well-maintained, complete, appropriate sprinkler system is not a magic wand. It requires the support of a well-considered integrated design for all the other elements of the building's fire protection. Unsatisfactory sprinkler performance can result from an inadequate water supply or faulty building construction. More broadly, unsatisfactory fire protection performance can occur if the building's design does not address all five elements of an integrated system—slowing the growth of

fire, automatic detection, automatic suppression, confining the fire, and occupant evacuation.[107]

Passive protection of structures and their occupants, including the use of asbestos in appropriate applications, became a much more elaborate science as the twentieth century progressed. Construction methods and details were specified in increasingly precise detail, and in dozens of fire safety applications these specifications included asbestos.

Fire-Resistive Materials in Ships

Like building fires, ship fires in the nineteenth century were an old problem for which both new dangers and new possible prevention, suppression, and containment strategies were emerging. The new dangers were brutally obvious: in the United States alone, more than two hundred steam vessels blew up between 1816 and 1846, prompting some of the earliest federal regulation of industry. Despite the efforts of Congress, multiple-fatality steam explosions on ships continued through the Civil War and Reconstruction periods.[108] While explosions gradually became less common as external combustion technology was mastered by its community of practice, ship and other maritime fires continued to cause significant loss of life well into the twentieth century.

Not only were ship power plants and the sparks from their stacks new sources of ignition in the nineteenth century, their fuel—coal—was itself a spontaneous ignition hazard, as were many other ships' cargoes.[109] Shipboard cooking had, of course, always been a fire hazard; by 1928 the Marine Committee of the NFPA was recommending the same asbestos board and metal protection for shipboard woodwork under cookstoves as for those in urban fire districts.[110] Cigarettes represented the same threat on a ship as on land, with the additional risk that had formerly been associated principally with carrying a pipe or a lantern into the hold: the ignition of vapors from cargo, fuel, or bilge. Ships of war, of course, carried ammunition that could "cook off" and explode in a fire. When electricity began to be used on ships, partly to reduce the risks of lighting with open flames, it was soon discovered that the two hundred miles or so of electrical wiring running through a passenger liner not only created a continuous stem-to-stern source of ignition, but the conduits through which the wiring ran proved ideal for the passage of smoke and flames.[111]

While more ships were made with metal hulls, bulkheads, and decks in the twentieth century, the vast majority retained their combustible interior finishes and furnishings until well after World War II.[112] Passenger liners were especially well provided with fuels of this type, including curtains, bedding, passenger clothing, and supplies of liquor. Passengers could not easily be disciplined, as sailors could, not to smoke in bed or set hot devices such as irons on wooden surfaces.

Just as the pandemic of theater fire mortality at the end of the nineteenth century directed public and professional attention to the dangers of public-assembly occupancy fires, a series of multiple-fatality marine fires between 1900 and 1935 brought investigation and action on ship fire-safety issues.

Ship fires, like theater fires, have their own literature of characteristic horrors. Ships are notoriously difficult to evacuate, especially in bad weather; cargoes can be combustible or even explosive. Combustion gases can build up in the hold or other enclosed areas and either explode lethally or asphyxiate persons who enter them. On ships with large open deck areas, such as ferries, passenger liners, and aircraft carriers, fire containment is difficult or impossible. If the ship's fuel tanks ignite, or if she carries highly combustible or explosive cargo, the entire vessel can explode and sink. If this should occur in port, other ships and onshore structures can be set afire and destroyed by burning brands or even by the explosion, as Halifax was in 1917 and Texas City in 1947. About 1,600 died in Halifax and 516 in Texas City.[113] To make the terror of ship fires complete, the water used to fight a fire cannot drain away as it does from a building on land and thus may sink or capsize the ship. Clearly, the best way to deal with ship fires, as with all others, is to prevent them from starting.

Not only the ships themselves were vulnerable to fire: in 1900, a fire at the Hoboken, New Jersey, pier of a German passenger line claimed 145 lives, most of them on burning ships, prompting investigation of how not only the ships but their loading and drydocking facilities could be made safer. After two more fires at this same pier, the NFPA's Committee on Piers and Wharves developed a model code in 1931 that included a provision that "side and end walls shall consists of non-combustible construction such as corrugated iron, corrugated molded asbestos, or other equivalent material."[114] Some port cities had already begun to incorporate such clauses into their codes; Seattle, for example, listed asbestos as a code-compliant waterfront roofing material in 1924.[115]

In June 1904, the *General Slocum*, an excursion steamboat operating in the East River and Long Island Sound, caught fire and burned with a full complement of families from St. Mark's Lutheran Church, a Lower East Side German immigrant congregation bound for a picnic on Long Island. Spread by a stiff breeze from its source of ignition in the ship's lamp room, the fire seized on the ship's superstructure and spread quickly up and down the forward stairways and through the large open areas of the decks, all of which were made of wood. The ship had no fire divisions; and when the captain turned the ship into the wind to try to make shore, the flames fanned back toward the stern, engulfing passengers who had fled there for safety. As passengers leapt from the ship, frantically attempting to save themselves and their children, they found that many of the cork life vests they had pulled from their storage nets had crumbled to dust in their canvas covers and were useless. The death toll was 1,021, the largest single-incident mortality in New York City until September 11, 2001.[116]

After the sinking of the *Titanic* in 1912, a group of industrial nations held a series of an international conferences in the following years concerned with developing standards for ship safety—the International Convention for the Safety of Life at Sea, usually abbreviated as SOLAS. Lessons learned during World War I and the early 1920s were incorporated into the second convention, SOLAS 1929, which specified standards and recommendations for the fire resistance of ships, including the use of asbestos in some applications. Ships were required to have fire-resistive bulkheads that could withstand a temperature of 815 degrees Celsius for an hour; asbestos was almost the only material light enough for a bulkhead that could pass this test.[117]

The importance of such standards, and of scrupulously revising, updating, and enforcing them in light of new developments, became increasingly apparent as ship fires in the 1930s relentlessly claimed lives on merchant and passenger vessels. Two French ships that had been built to the 1929 standards, the *Georges Philippar* and *L'Atlantique* burned with the loss of crew members in May 1932 (fifty deaths) and January 1933 (about twenty deaths) respectively, prompting the French government to forbid all use of wood in ship gangways, passages, decks, and living quarters unless firestopped by asbestos board.[118] *L'Atlantique* was almost certainly an electrical fire, although arson, as always, was considered; the findings of investigators after this disaster prompted a movement to upgrade wire insulation on ships and to install electrical cable of the proper low-iron grade of asbestos in marine applications.[119]

Both the British and the American navies had been using asbestos as boiler insulation since the late nineteenth century and were not slow to adopt it for interior finishes as well. By 1934 the international experts in marine construction were recommending asbestos as protection for woodwork near motor-exhaust piping and manifolds, for bulkheads and linings, for firestops separating wood from other materials, and for fire-resistive steel doors as well as the protection for wooden floors they had endorsed earlier.[120]

One of the most memorable maritime fires of the mid-thirties was that of the *Morro Castle* in September 1934 because it involved significant loss of life among the passengers rather than the crew and because of the bizarre circumstances of the disaster, including the death of the captain just seven hours before the fire. One hundred thirty-seven people died in this fire, which, like that of the *General Slocum*, occurred fairly close to shore and was spread in part by the turning of the ship into the wind by the interim captain, who had been hastily appointed after the death of his predecessor. Firefighting facilities were found to be woefully inadequate; and as in other cases I have described, large undivided fire areas and combustible interior finishes and furnishings allowed the blaze to spread rapidly.[121] Losses of ships to fire, with or without loss of life, were common before 1930 and much less so in the following decades after the introduction of fire-safety systems that included asbestos.[122]

The fire experiences of the *Georges Philippar*, *L'Atlantique*, and the *Morro Castle* were discussed and published in fire-safety professional journals of the late 1930s; but the updated standards developed from them did not appear in the text of SOLAS until the 1948–49 edition, by which time many additional grim and devastating ship fire-protection failure modes had been studied and documented.[123] In the meantime, marine insurance underwriters were able, to a limited extent, to enforce the lessons of the *Philippar* and *L'Atlantique* on shipbuilders by the simple expedient of refusing to insure them unless every feasible means of preventing, suppressing, and containing fire were employed, including covering wooden bulkheads with asbestos. Requirements became even stricter when the risk to shipping increased after hostilities began in 1939.[124] The 1949 SOLAS was more than twice the length of the 1929 edition; and like the *Fire Protection* and *Life Safety Handbooks*, every paragraph represented grief and loss: in the ten years after 1935, more human lives were in danger on the sea than in any previous period of human history.

4

Mass Destruction by Fire

Asbestos in World War II

Fire has been employed as a weapon since humankind first learned to control it; but not until large cities could be burned from the air did it become, as in the Second World War, a true weapon of mass destruction. Torching a village or a town is certainly destructive, and there will be fire casualties among the slow-moving—usually the very young, the sick, and the very old—but the procedure is laborious enough to give most of the population the opportunity to flee. A heavy rain will thwart the attack altogether. Most preindustrial villages consisted of single-story buildings from which escape was not difficult. Attempts to burn entire cities often came to grief: the attackers had to separate to carry the fire through the city and risked becoming lost, trapped by their own fires, or ambushed by the defenders. Moreover, there is a limit to how far a flaming projectile, such as a ball of pitch-soaked rags or a side of bacon, may be hurled even by a large catapult or cannon.[1] As components of military operations, large-scale urban conflagrations like those in Columbia, South Carolina, and Atlanta, Georgia, during the American Civil War were helped considerably by the local population, the weather, and sheer luck.[2]

The fires of World War II included several new elements that made them significantly different from those, for example, of Attila the Hun or Genghis Khan. First, incendiary materials were available that burned much hotter than the wood of torches or arrows dipped in pitch. Magnesium burns at about 3,600 degrees Fahrenheit and cannot be quenched in water, with which it reacts; thermite burns at just under 4,000 degrees Fahrenheit, which will melt most metals and ignite aluminum, which, in fact, is one of thermite's ingredients.[3]

Second, neither Attila the Hun nor Genghis Khan could fly. Military aviation made it possible to drop incendiaries into the heart of a city without sending troops into the streets. Reconnaissance from the air enabled attackers to locate areas of conurbation that would be especially vulnerable to fire. Third, the twentieth-century city housed stored fuels such as gasoline that could be targeted and used to spread flame to surrounding areas. Fires in some of these fuels are not readily extinguishable by traditional means, such as water; and in the case of those that are, high-explosive complements to incendiary bombing will not only smash structures literally into kindling but also break water mains, bring down power lines to impede firefighting, and make streets impassable for first responders.

Finally, the warriors of 1939–45 had acquired a set of skills and technologies that not only Attila and Genghis Khan but even Julius Caesar and Napoleon Bonaparte would have envied—that enabled them, within certain limits, to predict the weather and thus set fires that would spread faster than the adversary could react to or escape from. As warfare continued to be changed by technological innovations, the United States and its allies learned other effective methods of incinerating whole communities: the nuclear weapons used in Japan made horrific fires, and a few decades later we mastered the art of setting even very lush, moist Southeast Asian tropical vegetation alight with napalm, the incendiary material we had developed for use on Japanese cities two decades before.

This combination of new factors gave incendiary bombing the kind of advantage in the Second World War that machine guns and gas had had in the First: many of their dangers were new, and countermeasures were still in early stages of development. Moreover, these new elements applied to both land and naval warfare. Aircraft carriers, in 1939 a relatively new type of vessel, were and are essentially floating bombs carrying not only vast quantities of fuel and ordnance but thousands of sailors, each of whom was not only a precious commodity needed for the war effort but an individual intimately connected to a family and a constituency at home. In democracies, the waste of lives through carelessness or incompetence is not tolerated indefinitely, even in wartime.[4] Every effort had to be made, by all the combatant industrial democracies in World War II, to preserve both military and civilian lives while continuing to hurl destructive resources at the enemy. As British Air Chief Marshal John Slessor eloquently expressed it late in the war, "The most important social service a government can do for its people is to keep them alive and free."[5] Keeping them alive, as Slessor makes clear, came first; and preserving their lives against the fires of a global war meant the further development of fire prevention and suppression technology under emergency conditions.

The Global Market for Asbestos between the Wars

Scientific American described asbestos in September 1919 as a mineral with "new worlds to conquer," adding that "new uses for this material are being found almost daily" and endorsing it in every application from refractories to washing dishes. According to the magazine's editors, asbestos would even do windows.[6] In Canadian economist Stephen Leacock's 1911 science fiction short story "The Man in Asbestos," human beings in the thirtieth century dress exclusively in asbestos, a development resulting from "the revolt of woman and the fall of Fashion." During the First World War, many considerably more practical uses were found for asbestos, including protection of the notorious Elgin Marbles at the British Museum, which were covered in layers of it for the duration.[7] The mineral in fact received, during World War I, the ultimate compliment to its economic and military importance: it was officially classified as Absolute Contraband of War on October 14, 1915, and could not be legally exported from the United States after November 28, 1917. Even Canadian imports of asbestos to the United States were restricted. *Metallurgical and Chemical Engineering* described it in 1914 as a having been "transformed from a mineral curiosity into an industrial necessity."[8]

The U.S. and world markets for asbestos continued to grow through the twenties, with India, Australia, and a few other nations beginning to exploit their asbestos mineral resources and, in some cases, beginning manufacturing operations as well.[9] Although Henry W. Johns had pioneered asbestos as a roofing material in the late nineteenth century, his technology, which involved much hand swabbing of the roof surface with the asphalt-asbestos mix, was necessarily labor-intensive. In 1903, Dr. R. V. Mattison of Ambler, Pennsylvania, introduced asbestos-cement shingles for roofs and siding, invented in Italy a few years before, which moved rapidly into U.S., European, and Asian markets.

Asbestos-band brake linings, invented in England in 1896 and first manufactured in the United States ten years later, were tested in 1907 against linings of the competitor material, leather. After the leather was reduced to a cinder and the asbestos proved entirely unaffected, asbestos became almost overnight the only acceptable material for motor-vehicle brake linings, which were woven in the older band brakes and molded after 1924 when four-wheel internal brakes were introduced. As *Automotive Industries* noted in 1922, other applications of asbestos in automobiles included "gaskets, copper asbestos for manifold, cylinder heads, spark plug washers, wicks for exhaust pipes; paper and board asbestos around the exhaust pipe for heat insulation; pads for blanketing the radiator; sheets inside the hood, and many other uses."[10]

Glazed and textured asbestos pipe and wall tile were being manufactured in Italy and imported into the United States in 1920, and further efforts were made to make the material more appealing. The aesthetic limitations of a material available in a very limited range of colors had made asbestos building materials a hard sell to architects in any but the most mundane applications; but methods of adding new colors, contours, and textures to asbestos roofing, wallboard, and shingles helped open new markets in commercial and residential construction in the late twenties.[11] The light weight and ease of fastening of asbestos tile and roofing gave them considerable appeal in the do-it-yourself market of the thirties; architects and contractors appreciated the convenience and speed of assembly of fire-resistive asbestos-concrete wall units. The discovery that asbestos roofing could be applied directly over an older wooden-shingle roof covering made the material even more appealing.[12]

In 1931–32, demand was reduced by the Depression's effects on vehicle manufacture and construction. Even this did not slow growth for long, however, because by this time asbestos had many other uses besides these two. Laundry manuals advised the use of asbestos hot pads for irons and asbestos paper for commercial laundry mangles; asbestos was used in the cleaning and dressing of fur; the well-stocked home or restaurant kitchen had asbestos stove mats; home economists endorsed asbestos insulation, firestopping, and air-cell radiator linings for safety-conscious homemakers; and dyers used asbestos filters to extract and purify colors. The federal government experimented with balloons made of asbestos fabric at Scott Field, Illinois, in the 1920s.[13] Some theaters and schools kept asbestos blankets handy for extinguishing fires. In Britain, London Transport bought 9,715 sheets of "perforated asbestos wood" in 1938 for acoustical insulation of six and half miles of the Underground.[14] By 1939, Oliver Bowles could plausibly assert:

> Asbestos may be termed indispensable to modern life. As the chief constituent of brake-band linings and clutch facings it is essential to automotive transport; in the form of gaskets and packings it is a necessary part of steam-driven machinery; as a heat insulator it plays an important role in both household and factory construction and equipment; and combined with cement it is employed in the manufacture of vast quantities of roofing and other building materials.[15]

Dependent on other nations for 94 to 99 percent of its asbestos supplies before and during World War II, the United States was the world's largest importer of the mineral until the third quarter of the twentieth century.[16] It was also the world's largest exporter of asbestos products, the list of which grew as the century advanced. With no significant American production of the raw mineral, there was no tariff on the importation of raw asbestos, but early in the twentieth century there were significant trade barriers to protect U.S. asbestos manufacturing.[17] The two largest industrial uses of asbestos—manufacture of woven and later molded brake linings and building construction—fed a more general growth trend in asbestos demand. By 1941, U.S. manufacturers were shipping more than 80,000 squares of asbestos roofing to foreign customers, up from under 27,000 squares ten years before. As table 4.1 shows, exports of asbestos paper products and millboard had doubled in this period, shipments of textiles were up by almost 150 percent, and other manufactured asbestos products nearly quadrupled. Clutch facings of asbestos were first made in 1936,

TABLE 4.1.

Manufactured Asbestos Products Exported from the United States, 1931–64

Year	*Brake linings, not molded (linear feet)*	*Clutch facings and linings (number)*	*Paper, millboard, and roll board (short tons)*	*Pipe coverings and cement (short tons)*	*Textiles, yarn, and packing (short tons)*	*Magnesia and manufactures (short tons)*	*Asbestos roofing (squares)*	*Other manufactures (short tons)*
1931	3,791,500	[b]	635	1,050	636	1,453	26,556	1,196
1932	1,959,796	[b]	293	1,226	452	610	30,886	647
1933	1,651,425	[b]	439	910	568	695	85,532	839
1934	1,641,333	[b]	602	1,389	619	1,277	26,457	704
1935	1,426,520	[b]	767	1,233	715	922	31,141	921
1936	1,963,029	316,585	630	1,665	665	1,051	41,459	1,354
1937	1,633,558	499,870	869	2,384	762	1,567	37,026	1,889
1938	923,672	448,121	725	1,143	565	1,601	83,080	1,593
1939	886,069	326,493	819	2,213	891	1,483	54,634	2,315

1940	638,037	411,958	1,231	1,667	1,124	1,373	70,505	2,956
1941	1,277,562	100,6371	1,540	1,187	1,556	2,149	82,149	4,431
1942	749,449	1,431,512	679	1,063	1,017	1,305	97,668	3,000
1943	353,892	1,282,190	899	1,362	975	a	54,896	7,135
1944	321,019	1,412,600	773	1,738	1,096	1,266	59,580	5,353
1945	353,028	1,360,846	754	1,825	1,665	9,131	119,770	6,047
1946	740,670	1,196,241	653	1,103	2,327	17,423	129,728	8,734
1950	532,358	1,055,685	12,925[c]	1,143	1,215	a	a	a
1955	a	1,182,728	16,395[c]	3,040	1,210	a	a	a
1960	a	1,461,015	9,961	3,488	1,323	a	a	a
1964	a	2,046,247	12,988	3,951	1,520	a	a	a

Source: Minerals Yearbook.

[a]No data.

[b]No exports.

[c]Construction materials.

and by 1941 American manufacturers were selling more than 2.7 million of them abroad.[18]

Growth conditions in the industry in the early twentieth century created a seller's market for skilled labor in which unionization efforts could succeed. The insulation and construction side of American asbestos labor, nearly all male, organized early: as the Salamander Association of Boiler Felters and Insulators under the Knights of Labor in 1884 and later as the International Association of Heat and Frost Insulators and Asbestos Workers of the American Federation of Labor in 1903. Some plants on the textile side of asbestos manufacturing, in which women predominated, were represented during the 1940s and 1950s by the Textile Workers Union of America (Congress of Industrial Organizations [CIO]), but asbestos textile workers were never as heavily unionized as insulators were.

Most of the asbestos used in the United States came from Canada, the world's leading producer, or southern Africa, especially from what was then Rhodesia, the Transvaal, and the Union of South Africa.[19] The second largest producer, Soviet Russia, was the only nation mining significant quantities of the mineral that had its own asbestos-products industry, reactivating Peter the Great's eighteenth-century asbestos-manufacturing enterprise in the Urals in 1927. Instead of Peter's document bags and proximity clothing, however, the USSR's asbestos-products output in the early 1940s was mainly brake linings and other automotive components and boiler insulation for munitions and other factories.[20] Russia had been systematically mining asbestos since 1885 and was hoping to open up new sources in Siberia if adequate labor and transportation could be brought to the region.[21] The other major producing nations (fortunately for the later Allied cause, nearly all of them were British Commonwealth countries) were almost exclusively exporters of raw product.[22] Even more fortunate for the United States, two important sources of militarily significant types of asbestos, Canadian chrysotile and Bolivian crocidolite, were in our own hemisphere. The latter had been thought worthless until its utility for gas-mask filters was discovered.[23] World demand for asbestos was so high by the mid-thirties, even after the 1931–32 dip, that even marginal operations such as the ancient mines on Cyprus could produce and export asbestos profitably by 1935. Ominously in hindsight, nearly all the island's production in that year was sold to Japan, which was buying asbestos worldwide in record quantities.

Meanwhile, Germany was struggling desperately to produce synthetic asbestos and find natural deposits of significant size that would remain accessible in the event of war.[24] Both enterprises proved a failure. The synthetic amphibole reportedly produced by German scientist Rudolf Leutz in 1935 was found after the war to be a form of fiberglass, and the discovery of chrysotile in the Bayerischer Wald turned out to be an anticlimax because the deposit was too small to reduce Germany's dependence on foreign sources.[25] Germany imported 7,582 tons of raw fiber in 1932 and more than twice this quantity in 1934—20,154 tons, most of it from South Africa and Russia. Even after later gaining access to Italy's small supplies and after the conquests of Finland, Norway, and Czechoslovakia, which had minor deposits, Germany still could not meet her war needs for asbestos.[26]

The organization of the industry at this period was essentially vertical and concentrated in a relatively small number of manufacturing firms. The largest U.S. asbestos company, Johns-Manville Company, for example, owned a large proportion of the Canadian mines. Asbestos mines in Vermont and Arizona, insignificant as to both quantity and quality of product, were also under the control of U.S. capital, the former owned by the Ruberoid Corporation.[27] Turner & Newall in Britain owned many of the mines in South America and southern Africa, from which the company had been shipping asbestos to the United States since 1916. In 1933, the National Recovery Administration included the asbestos industry in its Codes of Fair Competition, which were apparently intended, somewhat paradoxically, to prevent both ruinous price wars and collusion to keep prices high in the several hundred industries the codes addressed.[28] The oligopolistic structure of the asbestos industry was noted by J. Hurstfield in 1944:

> Certain monopoly producers or groups of producers were able to exercise a considerable control over production and distribution by the very nature of their industrial structure, without needing to build up complicated and often transitory international cartels. This applied particularly to platinum, nickel, whale oil, wolfram, manganese ore, asbestos, camphor, quinine and, to a certain extent, oil.[29]

It is easy to see, in the world asbestos market of the second half of the 1930s, the foreshadowings of global warfare in the decade following. As noted, Germany and Japan, neither of which had their own sources of asbestos, were assiduously purchasing the material abroad for factories, aircraft, transport, and electrical uses; Soviet Russia was by 1935 accumulating a reserve larger than the entire annual world production at the time—18 million tons—while increasing its exports to Japan by twentyfold in the three years between 1931 and 1934. Russia was also building up its asbestos manufacturing industry, with assistance from American consultants and capital.[30] Japan, which had only very small deposits of the mineral in the home islands, had made exploitation of Manchurian asbestos in conquered Manchoukuo a priority by the end of the 1930s.[31] Italy was trying to bring its old and marginal asbestos sites back into production, as were other nations that saw opportunities in the rising demand for and prices of raw asbestos.[32] Most of Italy's output was shipped to Germany.

As Oliver Bowles, the U.S. Bureau of Mines' resident asbestos guru, remarked nearly every year of the 1930s in his annual report, any economic growth in any industrial country resulted in an increase in demand for asbestos because it had become essential to so many industries: conveyor rolls, belts, and facing for moving very hot materials such as glass; welding-booth curtains and blankets; paint spray booths; proximity clothing (see fig. 4.1); drying mats; diaphragms in electrolytic cells; dust collection bags in factories and bakeries; lehr (annealing furnace) curtains and cords used in glassmaking; belts for blue-print machines; oil and kerosene burner wicking; electrical tapes and tubing; gas burner mantles; wire-wiping cord; rope for sealing boilers; steam fittings; engines and gas generators; wire insulation; chemical filters; pad or wrapped insulation for marine and other turbines; coverings for the brine and ammonia pipes used in refrigerating equipment; adhesives; and electrical appliance insulation in both domestic and commercial applications, such as toasters, washing machines, and refrigerators. Some farming communities near asbestos mines in the 1920s even used the waste to neutralize soil acids and pave roads.[33]

Whenever goods or people moved, asbestos was needed for brake linings, locomotive firebox packing, automotive and truck firewalls, marine boiler insulation and fire-resistive bulkheads, to name only a few of the applications in

FIGURE 4.1 Proximity suits of asbestos worn by General Electric employees working at a steel annealing furnace. Note the slits for the lower leg and exposure of the shoes. Hood and gloves are separate from the gown. The eerie eyeglasses have lenses made of mica.

Photo from *Johns-Manville Service to the Iron and Steel Industry* (New York: Johns-Manville Company, 1930). Reprinted by permission.

which asbestos was used in transportation hardware. Johns-Manville's glossy catalog of products and services for railroads was 264 pages long in 1923, the one for power plants was 79 pages in 1931, and the catalog for the oil industry filled 219 pages the same year.[34] Even the recording and movie industries needed asbestos-containing acoustical tile, which was also becoming more widely used in other workplaces because, according to the medical authorities of the time, "the physical and mental energies of . . . employees were being sapped by the noise that surrounded them."[35]

By the mid-twenties, it was possible to order entire prefabricated fire-resistive buildings of asbestos-cement on steel framing that could be assembled with carpenters' tools, taken down, and reassembled somewhere else in a matter of hours. School buildings of this type could be purchased for $1,500 per room in 1926, and an entire bungalow with three rooms and a porch sold for $738. Altoona, Pennsylvania, Miami, Florida, and Salamanca, New York, all had "portable" school buildings of asbestos-cement construction. Small railroad stations could be completely fitted out with sheets of various kinds of asbestos, including the seats, counters, walls, desks, floors, and ceilings. Asbestos-lumber fencing was available to reduce combustion hazards from factory-chimney and locomotive sparks.[36] Asbestos-cement was considered well-nigh indispensable in worker housing, as a corrosion-preventive wrapping for iron petroleum pipe and as a fire-resistive substitute for metal in ductwork.[37] Asbestos fiber in a waterproof cement compound was used to patch roofs while enhancing their fire resistance; a similar compound was available for floors. Asbestos-protected metal was invaluable in docks, ductwork, and any application in which the metal would be exposed to fire, salt, refrigerants, extremes of temperature, corrosive fumes, sewage, acid, alkalis, sparks, smoke, electrolytic action, or condensation. Galvanized steel, for example, failed when exposed for long periods to "intense sulphurous gas." Johns-Manville sold a corrugated asbestos roof to a firm in Baltimore that needed a material that could withstand the fumes of decomposing bird guano; on a smaller scale, poultry farmers could solve the same problem by buying asbestos chicken-house kits by 1938.[38] In this decision environment, with so many kinds of enterprise dependent on asbestos, no developed nation could afford to neglect its supplies or lose sight of the means by which such supplies could be maintained in an emergency.

Moreover, asbestos types were not interchangeable. The supply had to be of a type suitable to the intended use of the material. Asbestos products used for fire resistance had to be certified "Underwriters' Grade" by the Underwriters' Laboratories and meet stringent performance and low-iron content standards to pass the UL tests for electrical uses. *Scientific American* noted in 1922 that "[t]he Navy has recently specified that asbestos insulated wire be utilized for the motors, etc. of a group of our newest submarines," explaining that the asbestos for this purpose must be "free from any traces of iron or other metallic oxides . . . lest its dielectric properties be impaired."[39] Beneficiation (the process of removing iron from raw asbestos) was rarely successful, despite strenuous experimental efforts by the U.S. Navy's research scientists and others.[40] South African chrysotile, for example, was well suited to electrical applications and high-acid environments because of its low iron content; in safety-critical end-uses, such as those on ships, there were then no acceptable substitutes. The Navy used it in packings, gaskets, turbine and electrical tape and insulation, and partitions.[41]

Only in certain kinds of insulation did government agencies see any possibility of substitution using fiberglass, and of these Graham Lee Moses of Westinghouse remarked in 1939 that "for the present it [fiberglass] should be applied with judgment only where its advantages serve a useful purpose" compared with asbestos, as in specialized types of electrical insulation. Where corrosive acids and alkalis were concerned, fiberglass could not be used, as there was no form of this material that would resist alkali until 1970. Engineer Brian Davis, writing in 1983, remarked that fiberglass "has been considered more brittle than asbestos, with inferior abrasion resistance."[42]

Transvaal amosite was also essential to the Navy, for no other material would meet its specifications for light weight and high insulation value. Fiberglass, which in any case weighed more than Navy standards would permit, was not as reliable as amosite at high temperatures, especially in the presence of acid. Canadian tests in 1941 showed that fiberglass tape, while stronger than asbestos before exposure to heat, had greatly inferior heat performance above 250 degrees Celsius (662 degrees Fahrenheit) and thus could not be used in applications where temperatures of 350–1,500 degrees Fahrenheit were routine.[43] For certain kinds of gas filters, including those used in military gas masks, there was no known substitute for the very fine fibers of Bolivian crocidolite.[44]

Although asbestos did not appear on the Harbord List of military raw materials published in 1921, less than two decades later, in late 1939—the asbestos industry's Diamond Jubilee year—with war in Europe already placing new stresses on American industry, the mineral was classified by the U.S. Army and Navy Munitions Board as a critical material, effective the first day of the following year. The National Defense Commission, correctly projecting an excess of military demand over available supply, recommended government procurement of a reserve in October 1940. Stockpiling of the required types and grades began the following year, on January 1, 1941. So urgent was the demand for asbestos that one member of Congress actually suggested exploiting very small U.S. deposits in Oregon, despite the absence of mechanized transportation out of the area, proposing that the mineral be brought out of the mountains on pack mules.[45]

Danger on the Sea: The New Ship Fires and the Old

The U.S. Navy, Coast Guard, and Maritime Commission (Merchant Marine) were the driving forces behind most of the asbestos conservation and stockpiling efforts coordinated by the War Production Board in the months after the United States entered World War II. After July 1943, the Combined Raw Materials Board, an Allied agency, allocated the available asbestos and other strategic and critical materials (SCM) between the two principal consumer nations of the war era, the United States and the United Kingdom.[46] Some metals, including nickel and tin, were on the "strategic" list, which outranked the "critical" list that included asbestos (see table 4.2).

The difference between these two lists was that strategic materials (for which the United States was dependent on foreign sources) were essential for fighting a war and posed significant challenges for procurement in wartime.

TABLE 4.2

Strategic and Critical Materials List, January 30, 1940

Strategic materials	*Critical materials*
Antimony	Aluminum
Chromium	Asbestos
Coconut-shell char	Cork
Manganese, ferrograde	Graphite
Manila fiber	Hides
Mercury	Iodine
Mica	Kapok
Nickel	Opium
Quartz crystal	Optical glass
Quinine	Phenol
Rubber	Platinum
Silk	Tanning materials
Tin	Toluol
Tungsten	Vanadium
	Wool

Source: U.S. War Production Board.

Critical materials were those "essential to national defense, the procurement problems of which in war would be less difficult than those of strategic materials." Asbestos was on the latter list because some types of asbestos were available domestically and from Canada.[47] As we have seen, there were no substitutes for asbestos in many of its important military—especially naval and aviation—applications; and it was itself a substitute for some materials urgently needed for defense purposes, including magnesia, which was used, ironically enough, to make the metallic magnesium used in the manufacture of some types of incendiary bombs as well as in the construction of refractories and the insulation of manufacturing equipment in the iron and steel industries.[48]

Delivery of some other military necessities such as petroleum, not included in the SCM list because U.S. sources were relatively secure, depended on methods of transport that used asbestos. More than 36,000 square feet of asbestos felt, for example, went into each mile of the Big Inch wartime emergency-oil pipeline's insulation; the Little Inch called for just over 30,000 square feet per mile.[49] As noted, trucks, trains, aircraft, and ships all used asbestos either for insulation or as a friction material or both. Demand for asbestos corrugated sheets, fire curtains, and other fire-resistive textiles to be used in construction of new manufacturing facilities exceeded supply by December 1941. Aircraft hangars and ordnance works, in particular, used large quantities of corrugated asbestos sheets and textiles because the structures had to be both quick to build and fire resistive in a decision environment in which nearly all kinds of metal were in short supply.[50] This was only one of many applications in which asbestos was used as a substitute for scarce and strategic metals; others included ductwork, gutters and downspouts, roofing, walls for prefabricated military structures, marine bulkheads, and flooring.[51]

These new demands on asbestos supplies obviously demanded cutbacks in traditional uses for the mineral. The first conservation order for asbestos was issued on January 20, 1942, just forty-four days after the attack on Pearl Harbor, restricting the uses of unmanufactured southern African asbestos to defense orders.[52] In February, the order was tightened and revised; and another order, M-123 was issued with additional restrictions. Within three months, 3,300 tons of African asbestos was lost at sea when the ship bringing it to the United States was attacked and sunk; M-123 was further revised on July 4 to include woven friction materials of asbestos and again in December 1942 with more detailed information about what could and could not be made from asbestos. By December 14, 1942, it was unlawful to manufacture any of the following from asbestos: theater curtains and scenery, vibration eliminators (except for implements of war), gun covers, radiator hose (again, except for those used in military hardware), firestops in non-military motor vehicles, conveyor belts except for those used in the glass industry (for which there was no substitute material), heaters and heater accessories other than those used by the military, filter sacks for liquids, parachute flare shields, and most kinds of clutch facing and brake linings for nonmilitary uses.[53] Although the order does not mention handcraft uses of asbestos, it is unlikely that materials such as asbestos-cement and putty were available to consumers for the construction of birdhouses, sculptured school projects, and similarly noncritical uses until after the war.[54]

In 1943, the U.S. government negotiated with the U.K. Board of Economic Warfare and British Ministry of Supply for the purchase of 57,490 short tons of African chrysotile, amosite, and crocidolite. During 1943 and 1944 additional restrictions were added to accommodate large increases in demand from the Navy, particularly for cable and lagging, and to permit the use in civilian applications of siding and roofing made from short-staple Canadian asbestos, which was recommended as protection against incendiary bombing.[55] In 1943, the Navy built 30,000 ships using about 912,000 shipyard workers and an astonishing 45,000 the following year with about 971,000 workers, building more than 3 million tons of warships in twelve months.[56] At these blazing speeds, the Navy's requirements were largely met by December 1944, so M-79 was amended to permit more uses of African asbestos. With the war in Europe concluding in the early months of 1945, additional restrictions were gradually dropped in a series of revisions; and all asbestos conservation orders were revoked by August 31, 1945.

The Navy's interest in asbestos was justified by its long experience with the mineral, dating from the 1870s. As noted in chapter 3, a ship fire is a serious matter. A ship fire in battle is even more serious for at least three reasons: (1) there is an armed enemy on hand making strenuous efforts both to extend the effects of the fire and to impede the efforts of the firefighters by killing or wounding them if possible; (2) the ship as a whole must be defended, diverting personnel from the work of firefighting; and (3) a ship of war carries explosives and other ordnance that can "cook off" and become hazards to the crew.[57] These dangers were bad enough on wooden ships, which could be and often were burned to the water line; but a modern ship, though more fire resistive in construction than a sailing ship, carries large quantities of fuel, which is by definition combustible. Electrical wiring, cooking and heating facilities, static electricity, and machine sparks provide plenty of sources of ignition in addition to those of battle.

If the vessel is an aircraft carrier, with not only the aircraft themselves but their fuels, coating and sealing compounds, paints and other finishes, and ordnance for both the aircraft and the ship itself, plus huge open deck spaces in which the aircraft can be stored, she is a deadly fire awaiting the torch of enemy action. The vulnerability of aircraft carriers was discovered very early after their introduction as naval vessels; and like all new construction technologies, the lessons of fire safety associated with them were learned the hard way, at the cost of hundreds of lives.[58]

The Pacific War in 1942 was in fact characterized by serious losses and damage to U.S. aircraft carriers in episodes in which fire played a significant role. Designed in the 1920s, the *Saratoga*-class carriers had many vulnerabilities, including wooden decks; and one, the *Lexington*, was lost to fire after enemy action at the Battle of the Coral Sea in May 1942, with a loss of 216 lives. By late 1942, the *Yorktown* had been sunk at Midway, the *Wasp* at Guadalcanal, and the *Saratoga* and the *Enterprise* were in repair facilities after battle damage, much of it from fire. The *Hornet* was lost to fire after a battle at the Santa Cruz Islands in October 1942, with a loss of 111 men. Fires had broken out all over the ship; and without power, fire-suppression capabilities were limited to buckets. Admiral James S. Russell, who had served on the *Yorktown* in 1939 and 1940, helped to design the new *Essex*-class and later carriers, which had greatly improved fire-safety systems, including some with asbestos fire curtains and others with asbestos-insulated metal doors that could be closed to divide the large, fire-prone hangar deck into separate fire areas.[59] These fire divisions can be clearly seen in the plan of the *Lexington* hangar deck in fig. 4.2. A rolled-up asbestos curtain is visible in the background in the *Yorktown* hangar bay, behind the aircraft, in fig. 4.3.

These precautions and others, such as "conflag" (conflagration) stations on hangar decks and sprinkler systems, helped reduce both the number of ships and their aircraft lost to fire after enemy action as well as the number of lives lost in carrier fires, until the kamikazes brought a new fire danger to American carriers in the last full year of the war. Admiral John S. Thach described the kamikaze strategy for producing deadly explosions and fires on carriers, and the accompanying American military debate, in October 1944:

> We were becoming quite concerned, of course, about this very effective method of hitting our carriers. It wouldn't have been worthwhile on any other target in the world. There's no use in diving into part of a factory, for example, or an antiaircraft installation. This was a weapon, for all practical purposes, far ahead of its time. It was actually a guided missile before we had any such things as guided missiles. It was guided by a human brain, human eyes and hands, and, even better than a guided missile, it could look, digest the information, change course, avoid damage, and get to the target.[60]

Interviewed about his World War II experiences long before September 11, 2001, Thach could not have known that even larger and more deadly human guided missiles, using much larger aircraft, were to dive on targets in New York City and Washington, D.C., setting them afire and killing their occupants, much as their predecessors had done more than a half century before. The carrier *Franklin* was hit by a kamikaze on March 19, 1945, with 725 men killed and 265 wounded; the ship was saved only by the heroic efforts of the surviving crew

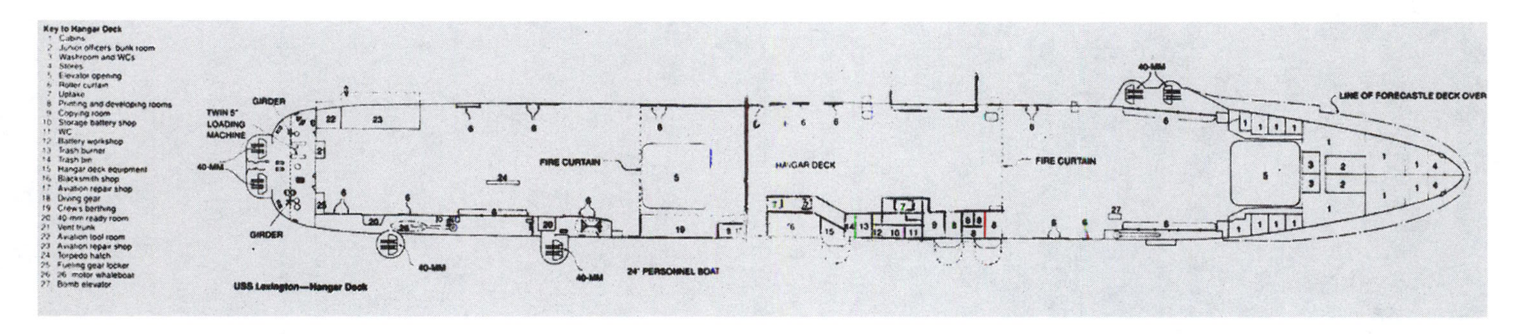

FIGURE 4.2 The U.S.S. *Lexington* was fitted with asbestos fire curtains at two points in her hangar deck, separating this very large space, in which fuels and explosives were routinely handled, into three fire divisions when lowered.

Alan Raven, *Essex-Class Carriers* (Annapolis, Md.: Naval Institute Press, 1988).

FIGURE 4.3 Hangar deck of the U.S.S. *Yorktown* during World War II. One of the asbestos curtains that separated the fire areas of the hangar can be seen rolled up against the ceiling at center right. The men at upper center are calmly watching a movie while ordnancemen prepare bombs in the foreground.

Photo courtesy of the Naval Historical Foundation.

and that of the nearby *Santa Fe*, which took off her wounded. Of the 379 bodies buried at sea from the *Franklin*, 210 were burned on all body surfaces, and 133 died of asphyxiation.[61] Bombs struck the ship in two of her three fire divisions, but these divisions protected the flight deck forward of the superstructure and the engine area, allowing both the evacuation of survivors and the eventual return of the ship to New York under her own power. The *Ommaney Bay* and the *Bismarck Sea* were also lost to fires after kamikaze attacks in 1945.[62] There were many fire losses among destroyers and other Navy ships as well, but the *Hornet* was the last major U.S. carrier to be sunk in World War II: the fire-safety system developed for *Essex*-class carriers, including the asbestos fire curtains (and, later, fire doors) and conflag stations, helped keep these fragile but indispensable ships of war afloat even after very severe damage by enemy action and fire.[63]

The improvement of fire safety on surface ships was well under way in the 1930s, as we have seen. By the time the Allied nations began building ships for war, asbestos-containing Marinite and similar products were the recommended materials for bulkheads and partitions, particularly those that formed the necessary fire divisions of the ship's interior spaces.[64] The mineral had become the recommended standard in a variety of shipboard applications in manuals of marine engineering, including use for high-temperature packings, gaskets, tape, electrical insulation, boiler casings, and walls and thermal insulation for pipes of all kinds.[65]

Asbestos proximity suits had been greatly improved for ease of donning and movement since General Electric's steel furnace workers were photographed by Johns-Manville in 1930, looking like representatives of the Spanish Inquisition suited up for an inspection of Hell (see fig. 4.4).[66] By 1942, the suits incorporated the relatively recent innovation of zippers and covered the entire worker, unlike the earlier robes, which had left areas of trouser leg and shoes exposed.[67] The new suits, first manufactured by the Mine Safety Appliances Company, incorporated a Bakelite helmet, heat-resistant glass visor, gloves, boots, and both front and rear aprons. Standards for the weight and type of asbestos cloth for use in these suits were also upgraded during the war and approved by the American Standards Association after testing by the National Bureau of Standards.

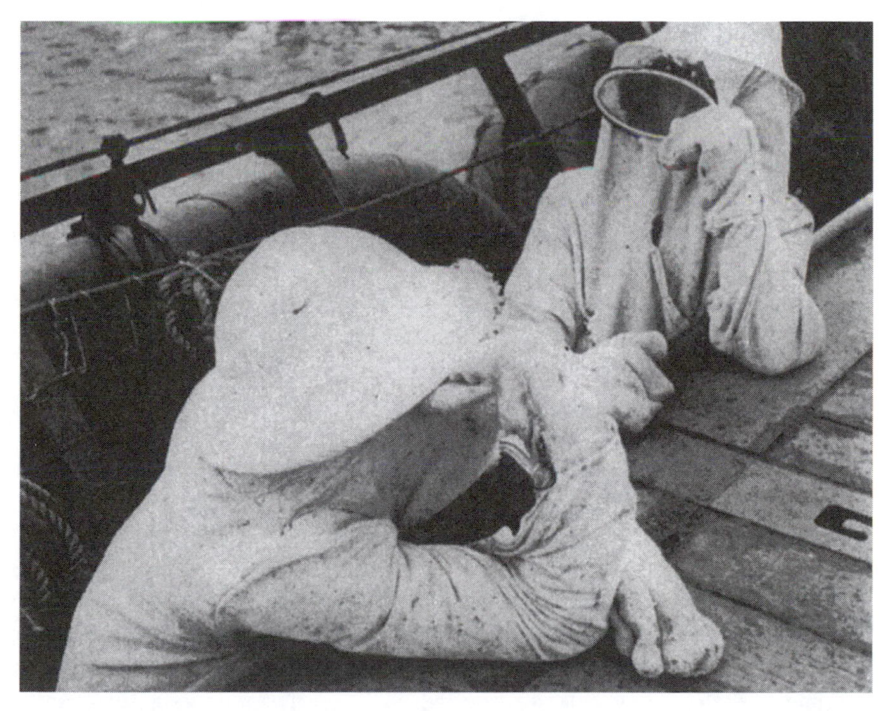

FIGURE 4.4 U.S. Navy photo of World War II firefighters in asbestos proximity suits. Note that the suits cover the entire body in a single garment; note also that the deck they lean against is made of wood. Suits of this type saved many lives on ships during the war.

Veterans of Foreign Wars, *Pictorial History of the Second World War*, vol. 1 (New York: Veterans of Foreign Wars, 1951). Reprinted by permission.

The Navy was a major customer for this type of safety clothing, equipping all shipboard firefighters with it as soon as production permitted. The U.S. Army ordered a million yards of asbestos cloth by August 1945, enough "to make complete fireproof fire-fighting suits for fully five divisions of men—seventy-five thousand G.I.s."[68] These garments, known in the Royal Navy as "fearnought suits," weighed "only" thirteen pounds; and they and their successors, which used reflective metallic cloth as well as asbestos, proved invaluable for rescuing pilots from burning aircraft. Tests conducted between 1933 and 1936 showed that a person wearing an asbestos suit could survive temperatures of 1,700 degrees Fahrenheit.[69] Postwar experiments with fiberglass showed that while this material had the desirable quality of low thermal conductivity, "[g]lass fabrics can withstand temperatures of 1100°F or more, and asbestos 2100°F or more," an impressive thousand-degree advantage of asbestos over fiberglass. This margin is especially significant in the case of metal fires, which are one of the most dangerous types of shipboard fire.[70]

Mineral wool was tried and found wanting during the war for the protection of some types of ship cargo, as in the packaging of battery electrolytes for shipment to Britain. The U.S. War Shipping Administration had initially packed one-gallon glass bottles of sulphuric acid in mineral-wool batts set into fiberboard containers, which were in turn set into wooden boxes for maximum protection. L. W. Bauer wrote in 1945:

> This method was used for about a year, until a flood of complaints began coming in from overseas. The British Ministry of Supply, the [U.S.] Coast Guard and the War Shipping Administration urged strongly that a pack be developed that would completely absorb the contents in case of breakage or leakage to prevent damage to other cargo and to guard against injury to personnel.
>
> A working committee appointed by the Joint Army-Navy Board investigated. It was found that the mineral wool batts would absorb only a small portion of the gallon bottle of electrolyte. Chemical action would set in, causing the mineral wool to throw off obnoxious fumes and heat up to a dangerous degree.

Further tests showed that seven pounds of asbestos would absorb all the acid from a gallon bottle without heating above 95 degrees Fahrenheit, so eleven pounds of the material were included in the packing of all subsequent shipments.[71]

Ships on Fire after 1945

Although the United States and its allies learned much about the prevention and suppression of ship fires during the war, the need for fire-resistive marine construction remained obvious. After the war there was another ship disaster, that of the passenger liner *Noronic* at about midnight on September 17, 1949, while she was tied up at a pier in Toronto. A small fire in a linen closet became a raging bonfire that claimed the lives of 118 of 695 persons on board because the ship had no fire divisions; the wooden interior finishes, flammable furnishings, and combustible partitions between passenger cabins spread the fire and burned so intensely that the NFPA investigator did not think it would have been possible to launch lifeboats had the vessel been at sea. While asbestos-containing and other types of incombustible fire divisions and partitions had been standard on most ocean-going ships since the 1930s, passenger shipping on the Great Lakes in 1949 was not required to conform to SOLAS standards for fire safety.[72] Neither the *Noronic* nor her sister ship, the *Hamonic*, did so; the latter was lost to fire off Point Edward, Ontario, on July 17, 1945, fortunately only after the rescue of the passengers and crew. Four years later, the *Noronic* was not so lucky.[73]

The military was not exempt, either. On May 26, 1954, an explosion and fire on the aircraft carrier *Bennington*, then off Quonset Point, Rhode Island, killed 103 crew members. During the Vietnam war, there were two disastrous aircraft carrier fires, neither the result of enemy action. Even with the improvements to carrier fire safety established during the Second World War, including fire divisions between hangar bays and conflag stations in each, on October 26, 1966, the *Essex*-class carrier *Oriskany* was set afire by the mishandling of pyrotechnic flares, with the loss of forty-four lives. The *Forrestal*, a supercarrier built in 1955 and namesake of a new class of ship, suffered the loss of 134 men when an accidentally fired rocket ignited the gas tank of future senator John McCain's aircraft, setting fire to other aircraft and their fuels. Several of the World War II–era thousand-pound bombs being loaded into the planes cooked off and exploded, killing most of the ship's specialized and highly trained firefighting crew within the first few minutes of the fire.[74]

Fire-resistance experiments on both structural materials and finishes were performed by the Navy's research branches throughout the Korean and Vietnam war eras, in which asbestos continued to show its advantages over other materials. Even after the public controversy over the use of asbestos began in 1964–65, the Navy continued to stand by its test results and try to protect the lives of its service members by making their environments as fire resistive as possible.[75]

The U.S. Navy's concern, and its reluctance to abandon the use of asbestos in ships, appears to have been justified. During the Falklands War, the failure, melting, and subsequent ignition of the aluminum bulkheads installed on British Navy ships as substitutes for asbestos-containing fire divisions caused a larger percentage of burn casualties, on a per capita basis, than the service had experienced during World War II.[76] On the U.S.S. *Stark* in May 1987, during the first Gulf War, 37 of 220 men were killed when an Iraqi Mirage hit the ship with two Exocet missiles. Melting or burning aluminum bulkheads caused similar injuries to those seen on the British Navy ships in the Falklands War, as sailors struggled to extinguish this very stubborn type of metal fire. Most of the *Stark*'s casualties were from burns, blast, and smoke inhalation.[77]

Letters of Fire: Incendiary Bombing in Europe and Japan

Incendiary bombing was first used in the Second World War in the summer of 1940, part of the Luftwaffe's effort to undermine the will of the British to resist what was then thought to be an imminent German invasion across the channel. The British responded with justifiably belligerent indignation rather than demoralization. They reorganized their fire services into one large national organization and trained civilian groups into effective local fire spotting and suppression teams. Asbestos played a significant role in both military and civilian responses to the German threat. Like other military aviation arms of the period, the Royal Air Force (RAF) used asbestos for both aircraft engine insulation and the proximity clothing used in fighting aircraft and airfield fires.[78] Temporary structures, including housing, were roofed with asbestos to save scarce metal and protect against fire; asbestos curtains were sometimes used as door coverings for air-raid shelters. Asbestos blankets were dropped over incendiary and other small fires to smother them. Gloves and aprons of asbestos were recommended to the teams, most of them composed of civilians, who disposed of incendiary bombs by, for example, dropping them into their bathtubs, which had been left filled after the last use for this purpose. Another British method of extinguishing incendiaries impressed American fire engineer Horatio Bond when he visited England in 1940: the use of Bestobells, a type of bomb snuffer made of asbestos over a wire frame attached to a six-foot handle with a hook through a metal ring. The device looked like a cross between a candle snuffer and a butterfly net and was reported to be very effective (see fig. 4.5). Instructions to the British citizenry for disposing of incendiaries and for other civilian defense tasks were set forth in a widely distributed government pamphlet with the characteristically understated title, *Objects Dropped from the Sky*.[79] One of the most significant aspects of the British response to the Blitz, however, was the RAF retaliation against German cities, in which American fire engineers were eventually to play an important part.

It was not until a year after the United States entered the war that fire-protection engineers became directly involved in the planning and evaluation of strategic incendiary bombing. After the war, the NFPA's chief engineer, James K. McElroy, described it as a "hard sell" to persuade the British Incendiary Bomb Tests panel in early 1944 that preparing fire-vulnerability maps and target analysis from aerial reconnaissance photographs was a worthwhile activity. These

FIGURE 4.5 A team of civilian fire-suppression volunteers in London trains in 1941 with Bestobells, wire-mesh domes sprayed with asbestos used to extinguish German incendiary bombs. American fire engineers reported that these devices were quite effective.

Photo by British Combine for Bestobell, Ltd. Reprinted by permission.

maps were essentially mirror images of the fire-risk maps prepared by the Sanborn Map Company for the insurance industry (see fig. 4.6). McElroy, Horatio Bond (also of the NFPA), and Major Forrest J. Sanborn, formerly of Factory Mutual Laboratories and Improved Risk Mutuals and by 1942 an officer in the U.S. Chemical Warfare Service, spent the better part of 1944 studying aerial photographs and maps of German cities. By late 1944 and early 1945, they were

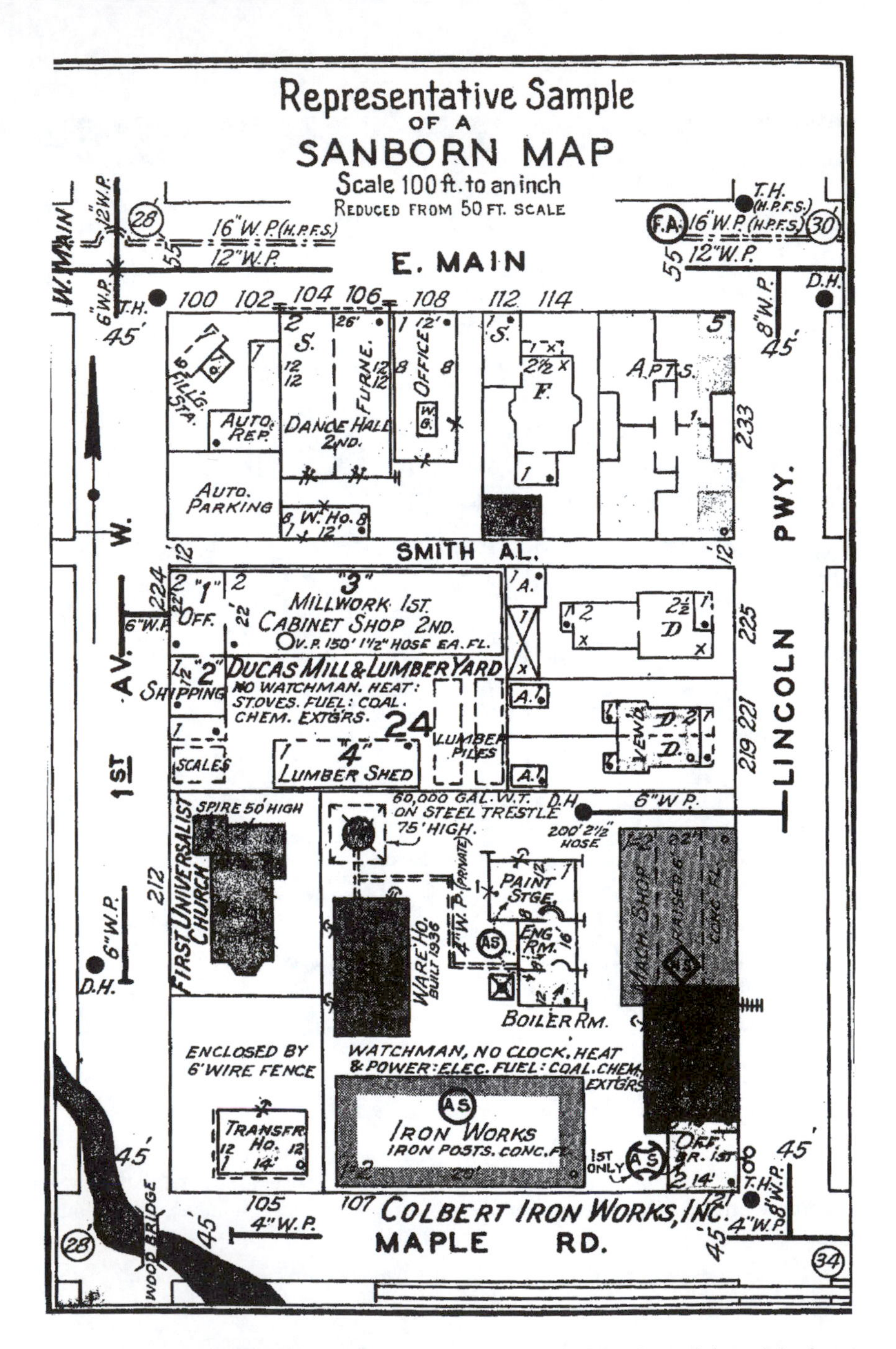

FIGURE 4.6 Sample Sanborn fire-insurance map, 1948. The colors of the original were coded to show whether or not the structures depicted were made of fire-resistive materials. Metal roofing, for example, was gray, as was asbestos, although in the latter case the letters "AS" appeared in a circle. The fire-vulnerability maps drawn by underwriters working for the military during World War II were essentially negative images of American fire-resistance maps.

Copyright © 2001 The Sanborn Map Company, Sanborn Library, LLC.

looking at such documents for Japan as well and expecting similar difficulties in persuading commanders in the field to use their analyses. McElroy relates with evident satisfaction his initial approach to the War Department in August 1944:

> After all our struggles in the European theatre to obtain acceptance of our idea of maps and target analysis, there came the day when we were asked to go in to General Lauris Norstadt's office in the Pentagon with our first Japanese industrial target map. We had the map hand colored. The reinforced concrete structures were shown in brown [a modified form of the convention of American Sanborn insurance maps], the combustible roof part of the target in red, and so on. We went in and laid it on his desk. That young fellow, who was Deputy to General Arnold in command of the Twentieth Air Force, wanted to send our only hand-colored sheet of paper right out to General LeMay in the China-Burma-India Theatre of Operations that afternoon. It was such a different reaction from what we had had in the early days in England that I was pretty thoroughly excited when I came out of his office.[80]

Even without the assistance of American fire-safety professionals, the RAF had succeeded in wreaking significant havoc on German cities between 1940 and 1943. Their own fire authorities were much too busy coping with the effects of the Blitz to be able to devote time to planning the destruction of cities elsewhere, and there was some initial resistance in Britain to the idea of using fire professionals as part of a military operation.

The *Strategic Bombing Survey* documented the destruction of Germany in text and pictures drawn from the stereographic photography done as part of the reconnaissance and bombing, contemporaneous pictures from the ground from German sources captured at the end of the war, and photodocumentation during the postwar occupation of Germany. Written and oral surveys and interviews, including many with firefighters and other civilian defense planners, helped complete a picture of mass destruction beyond the capabilities of any existing means of preventing and fighting fire. The addition of American forces and American fire-safety professionals had the desirable effect, from the Allied point of view, of making the attacks more deadly and destructive with less bomb tonnage and fewer Allied lives lost. The attacks had several purposes: destruction of industrial capacity, which proved difficult to do by this means; disruption of transportation of goods, troops, and people; promotion of absenteeism from work on the part of the "dehoused"; and demoralization of the nation waging war.

In the latter three instances, especially the last, the bombing campaign seems to have been more successful than in its first goal of bringing war production to a halt. One of the consequences, apparently not foreseen by Allied planners (but welcomed nevertheless), was that considerable labor was diverted from other industry into the reconstruction of buildings destroyed or damaged by air raids. Early in the war, German mobilization authorities had decided that carpenters, roofers, electricians, glaziers, and so on were not sufficiently important to the war economy to make them draft-exempt or even deferrable, so few were left to rebuild when their communities burned down.

And burn down they did. Between 1939 and 1945, according to the *Strategic Bombing Survey*, an estimated 2.15 million homes were destroyed, 11 percent of Germany's total housing, rendering more than 7.5 million persons homeless. Officially, more than 300,000 people died, a third of them in Hamburg and Dresden alone.[81] Some postwar estimates have put these figures much higher. This was a remarkable record of fire destruction in a nation with comparatively strict urban building codes. German buildings had no wooden porches, outbuildings, or fences to spread fire and had a fairly well developed system of planning for civilian shelters, although shortages of labor and materials brought most of the plans involving new construction to a standstill.[82]

Hamburg, for example, had very few frame structures and virtually no combustible interior walls in buildings; its wharves were all of fire-resistive construction, a fact noted with admiration by the Allied professionals on the survey team who had fire experience in American port facilities.[83] The city did, however, have some wooden roofs and floors; and nearly all buildings, even those of "massive" (slow-burning) construction, had wooden roof trusses that proved to be the Achilles heel of the Hamburg built environment. After the fire, U.S. Strategic Bombing Survey photographs showed roofless shells of buildings, completely burned out, facing streets that functioned like horizontal chimneys during the firestorm of August 3, 1943, described as "the worst holocaust in history" because the burning of Tokyo was at that time still almost two years in the future. The Strategic Bombing Survey's 1947 *Hamburg Field Report* made no attempt to sugar-coat the catastrophe:

> *Progress of Fire*: The tremendous hurricane of fire caused the air to be drawn toward the fire from all directions with such terrific velocity that it tore trees apart, and prevented firemen from coming within range with hose streams. Soon it caused building walls to collapse into streets, and further prevented the department from bringing apparatus into the area. In some cases it prevented them from withdrawing equipment, which, as a result, was destroyed, and, in some instances, firemen were killed. Because of the terrific heat and showers of embers, existing open spaces, even parks, could not be used by the department, and they eventually discontinued their efforts to extinguish or hold the fire, using every means to rescue the thousands who were trapped in basement shelters in the area. The fire department estimates that fifty-five thousand (55,000) persons lost their lives. Thousands of victims, later found

> in basement shelters, showed no indication of having been burned, but apparently they died from lack of oxygen, heat inhalation, or asphyxiation from coal or other gases. Other thousands in attempting to escape from shelters down flame-swept streets were burned to death. Those missing are probably buried in the ruins, or their bodies have been completely consumed by fire. Of the number cited, forty-five thousand (45,000) bodies were recovered. The department reports that, through their efforts, eighteen thousand (18,000) persons were rescued from the fire area. The damage to property was enormous; thirty-five thousand, seven hundred (35,700) dwelling apartments were totally destroyed; four thousand six hundred and sixty (4,660) severely damaged. From a total of four hundred fifty thousand, eight hundred (450,800) family apartments, two hundred fifty-three thousand, four hundred (253,400) were destroyed or made unfit for use. A total of five and nine-tenths (5.9) square miles of buildings was totally destroyed. Many industrial establishments, stores, ships and automobiles were destroyed or severely damaged.[84]

Postwar estimates of the number of dead in Hamburg place the figure closer to 60,000, and some German sources offer even higher estimates. Dresden, too, was devastated, with more than 30,000 people dead and about 80 percent of its structures destroyed or damaged.[85]

But even these massive disasters were exceeded in destructive scope by the incendiary attacks on Japan, begun late in 1944 with the newly developed B-29 bomber and, as we have seen, the assistance of American fire-safety professionals, particularly underwriters who could draw fire-vulnerability maps from reconnaissance photographs. Japan's built environment had almost no fire resistance; and its fire services struggled with obsolete equipment and little in the way of infrastructure, such as accessible water supplies, to bring to bear against an overwhelming incendiary onslaught against nearly every Japanese city larger than Dayton, Ohio.

Writing of Japan's war experience retrospectively in 1949, Strategic Bombing Survey economist Jerome Bernard Cohen observed, "Suzuki-san started with little and ended with less."[86] This was true of all the necessities of life, but the built environment in particular suffered. The Strategic Bombing Survey's Physical Damage Division reported after the war that most Japanese residences had been made of wood with incombustible tile roofs, but most burned to the ground within ten minutes when exposed to fire from the sides. Burning brands from above were not a hazard, but contents and walls ignited through unprotected windows, causing wooden roof decks to collapse from within. Factories were typically made of steel, iron, or wood-frame construction, with sawtooth roofs covered with asbestos or iron roofing.

Photographs of the results of strategic bombing in Japanese cities show post-incendiary moonscapes from which the burned-out hulks of fire-resistive buildings jut at wide intervals, like rock outcroppings on a dead planet. Not only was the Japanese built environment an almost ideal target for fire, its population had the added disadvantages both of being inadequately equipped and trained by their government to deal with incendiary bombing and of having been told they would never need this equipment or training because the Japanese homeland was secure. Robert Guillain, a French journalist interned in Japan during the war, was shocked by the official neglect of civilian defense and horrified by the results of incendiary bombing. He described the post-raid urban environment as "a sort of underbrush of roasted sheet-metal and reddish-blackish iron and heavy shards of gray tile! It was almost impossible to see where streets and alleys had run between the burned-out squares that once were houses."[87] During bombing raids, Japanese civilians stood outside, many of them on roofs, watching for the fall of incendiaries and struggling to extinguish them with sand and the rush mats, soaked with water, that in ordinary times were used as floor coverings.

The incendiary raid of March 9, 1945, destroyed 85 of Tokyo's 210 square miles of cityscape, killing an estimated 130,000 persons. The official count was 83,793, but the Japanese considered this figure conservative. Flames started by the incendiaries made the central portion of the capital of Japan into what historian Martin Caidin later called "a monstrous, diseased flatness." The fire was out of fire-service control within thirty minutes of the first fall of bombs. The Strategic Bombing Survey later said that "the fire susceptibility of Zone 1 [central Tokyo] was probably greater than of any other similar area in the world." Discussing the period in which the fire raced through the city at an estimated fifty miles per hour, the survey continued: "Probably more persons lost their lives by fire at Tokyo in a 6-hour period than at any time in the history of man."[88] Strategic Bombing surveyor Karl Compton told NFPA members in 1946, "There were places in the middle of Tokyo where one could stand and look for five miles and not see anything standing except an occasional masonry wall, perhaps the end of some masonry building."[89] Cities elsewhere in Japan were burned to cinders one by one between March and August of that year; in Toyama, a city with a population of 127,000, 97 percent of the structures were totally destroyed.

Assessing the condition of Japan after four months of incendiary bombing and two days of atomic attack, members of the Strategic Bombing Survey made very detailed observations of eight cities between October and December of 1945. Survey teams consisted of about twenty persons, of whom at least one was a fire-safety professional, and most teams had two. Other members included draftspersons, structural engineers, ordnance experts, photographers, clerks, and interpreters. Seeking lessons for both possible future incendiary warfare and, perhaps more important, countermeasures against incendiary and nuclear attack, the surveyors recorded the performance of various types of building construction under fire and atomic explosion conditions. Photographs of the destruction showed many industrial buildings bare of sheathing but with repairable framing, some with machinery inside that had survived the fire but had suffered from the action of weather after destruction of the roof. The teams were impressed by the performance of asbestos in both contexts; although sheathing and roofing cracked and usually fell off when exposed to very high temperatures, "[n]o cases were noted where exposure fires cracked the asbestos and exposed the contents to fire." The same definitely could not be said of the wood and earth structures in which most Japanese lived. At Nagasaki, team members noted:

> Steel-frame buildings covered with corrugated asbestos siding and roofing generally suffered less structural damage than those buildings having corrugated iron or sheet-metal covering. The blast effects immediately ripped off or crumbled the asbestos material, leaving no wide surfaces exposed against which damaging blast pressures could be exerted and transmitted to the framework. Metal siding, however, transferred pressure to the structural members, causing distortion and general collapse.[90]

This observation was stressed in their conclusions about structural considerations of civilian defense against nuclear attack, a judgment that was to have important implications in the next two decades.[91]

Fire on the U.S. Home Front, 1939–1946

The six years of World War II, in which the United States was a combatant nation for three, brought into American homes frightening images of the destructive power of fire used as a weapon. Because some components of the fire-safety system established between 1896 and 1939 were reduced in enforcement and implementation priority by the demands of the war, including the recruitment of many fire inspectors and engineers into the war effort, the war years were the twentieth-century peak of major loss-of-life building fires in the United States (see fig. 1.2). Another triple-digit multiple-fatality fire occurred just after the end of the war, reinforcing the lesson that home dangers do not go away when we are threatened from abroad, nor are they significantly reduced in a secure homeland after armistice but before resumption of peacetime safety controls on the built environment. Americans once again learned that only professionally designed codes, well-planned fire-safety systems, and vigorous enforcement protect civilians and service members at home, on leave, or after demobilization.

There were a variety of interrelated reasons for the large World War II spike in multiple-fatality building fires. As military theorist Carl von Clausewitz observed in his 1833 *On War*, "Everything in war is very simple, but the simplest thing is difficult." The priorities were obvious, but it was much less clear how to order them in such a way as to project power into both the European and the Pacific theaters of war, supply our own forces and those of our allies with war production, and maintain security on the home front, including protecting human and other resources from fire. Factories suddenly on three-shift operation in temporary or newly built structures, or in old ones after years of deferred maintenance from the Depression years, could not get certain kinds of building materials, such as metal roofing, which were, of course, reserved for the military. Copper and other metals for sprinkler systems were virtually unobtainable, even had there been available labor to install and maintain them. Repairs to existing fire-safety systems were deferred due to labor shortages and the urgent needs of immediate production. Goods in warehouses were piled so high in some cases that sprinklers were blocked and rendered entirely inoperable. Fire-safety engineers pleaded desperately that cutting safety corners in this way was false economy and cited cases such as the bonfire in January 1942, which involved enough stored wheat in Superior, Wisconsin, to provide "an army of approximately 700,000 men" with bread for a year.[92]

Workers brought in to work in new plants, new production operations, and extra shifts had to be housed; and the same construction limitations applied. People, machinery, and materials were crowded together with little regard for prewar building, exit, and occupancy codes. Fire-resistive materials, including asbestos, were almost impossible to obtain except for defense end-uses. This was true of leisure as well as work; production workers and their families, with money to spend for the first time in more than a decade, as well as service members home on leave, packed bars, nightclubs, movie theaters, circuses, and other places of entertainment. The one bright spot in this gloomy picture of home front disregard for fire safety was the situation of piers and wharves. The Allies had learned an important lesson in port fire safety during the First World War, when more than 1,600 civilians and military personnel were killed in Halifax, Nova Scotia, on December 6, 1917, after a French ship loaded with ammunition was rammed by another vessel and exploded. A substantial number of the casualties in this disaster resulted from the fire that consumed most of the town after the explosion.[93]

In the Second World War, any U.S. waterfront on which defense materiel was loaded was placed under the jurisdiction of the Coast Guard, which imposed much stricter codes and standards for fire safety than most local municipalities had done, requiring asbestos firestopping of wooden piers, for example, to prevent the spread of fires like the one that had consumed a large portion of Hoboken's marine infrastructure in June 1900. Additional major fires without loss of life had occurred at Hoboken Pier in 1927 and 1930. This remarkably fireprone pier experienced its fourth major fire in August 1944, caused by a nitrocellulose explosion, but the fire safety system then in place prevented loss of life.[94]

Regulations for fire prevention in ships in layup for repairs and modification were revised and more vigorously enforced by the Coast Guard and the Navy after the loss of the French liner *Normandie*, rechristened U.S.S. *Lafayette*, at New York City's Pier 88 in the North River on the evening of February 9, 1942. The *Lafayette* was to be adaptively reused as a troop transport. The Navy Department Court of Inquiry later determined that the fire started when a welder at work in the Grand Lounge neglected to set up the required asbestos curtain between his welding torch and 1,100 bales of life preservers. Once these ignited, all the furnishings and the carpet were soon aflame. By 2:30 the next morning, the ship had taken a seventy-nine-degree list from the water used for firefighting and was lying on her port side. One person was dead and 285 were injured, many of them firefighters. The ship was cut up for scrap after the war, a $55 million loss to the American taxpayer.[95]

The other major factor contributing to the high death rate from fires in the United States in the 1940s is that there was hardly anybody minding the store. Firefighters, fire-safety engineers, building inspectors, and even insurance underwriters either entered the armed services or were mobilized for national service in other ways. As we have seen, scores of fire engineers and insurance underwriters were involved in the planning of strategic bombing in Germany and Japan. Many firefighters went to Britain to assist with fire suppression during the Blitz; and large stocks of American pipe, hose, and other equipment were sent to London in 1939 and 1940. What property insurance underwriters were left in civilian life were kept frenetically busy inspecting defense plants and other War Department installations.[96] By May 1944, 112 members of the National Board of Fire Underwriters, working on a "dollar-a-year" basis, according to the U.S. Navy, "surveyed 431 properties in 25 states. These included 202 naval shore establishments, 197 shipyards, three hotels and 29 schools."[97] Obviously, little time remained for inspections of such low-priority venues as nightclubs.

The remaining firefighters in the municipal services struggled along with insufficient labor and often obsolete equipment as well. The Depression had, of course, cut into local budgets in most cases, preventing the purchasing of new fire equipment.[98] After December 7, 1941, new firefighting vehicles and equipment became very difficult to obtain for civilian purposes. All of these issues combined to make the U.S. fire environment much more risky than it had been before and would be later.

The first major World War II–era multiple-fatality fire in the United States attracted little publicity at the time and is the only one of the catastrophic fires

I am about to address that has not been the subject of a book-length history.[99] This was the fire at the Rhythm Club, in Natchez, Mississippi, on April 23, 1940, in which 207 members of an audience of 746 African Americans died in a metal-and-frame structure with only one exit that had been decorated with Spanish moss for a special evening of music and dancing. A hamburger stand near the only door ignited the moss, which set fire to the structure's wooden floors and wainscoting and, as the flaming decorations fell, to the clothing of the audience. The windows had been boarded up and nailed shut; the doors opened inward. Apparently building codes were not enforced in the African American part of the community, if indeed they were enforced anywhere in Natchez at that time.[100] A brief (six-page) technical investigation of the fire was published in the *NFPA Quarterly* in 1940, but the disaster did not make national headlines outside the fire-safety community.[101]

The second great home-front fire occurred after the United States entered the war; and tragically, a number of the nearly five hundred deaths were those of American service members: fifty-one men and two women. The Cocoanut Grove nightclub fire of November 28, 1942, was America's second most deadly single-building fire, after the Iroquois Theatre, a disaster with which it had many elements in common.[102]

Like the Iroquois, the Cocoanut Grove was a place of entertainment—what would now be called a public-assembly occupancy—although according to Boston's building laws in 1942, restaurants, bars, and nightclubs were not considered places of public assembly and did not have to conform to the same standards as those establishments. Like the Iroquois, the Grove had been inspected by local officials, in this case only eight days before the fire, but had been passed as "good" despite the permanently locked condition of some of its exits. Only two were unlocked and available to guests at the start of the fire, one of which was a revolving door, the other of which opened inward. Thus, like the Iroquois, the Grove fire became notorious for its large piles of bodies at impassable exits: about a hundred at the door that opened inward and two hundred at the revolving doors.

Both buildings were structurally fire resistive and sustained very little permanent damage but had fatally combustible interior finishes and contents. The Grove had cloth drop ceilings and leatherette wall coverings, in addition to flimsy straw-like palm tree decorations and ceiling ornaments. These were reportedly tested with a match by the building inspector on November 20 and failed to ignite. In both fires the lights went out at a critical point in the rapid spread of the fire, and a flashover (sudden ignition of all combustible materials) sent a rolling wave of flame into the crowd, some members of which died so quickly that they did not have time to leave their seats. One Grove victim died in her chair, with her hand still on her cocktail glass. Invisible or locked exits were factors in both fires, as were confusing room layouts. In neither establishment had the staff received any fire drills or training. Shady dealings between the city and the management were suspected in both cases; certainly there was official negligence on the part of the Chicago and Boston building authorities, even if evidence of outright corruption was discounted, as other historians, especially Edward Keyes, have pointed out.

Both houses were packed to more than capacity; the Grove's certificate of occupancy was for six hundred persons, but more than a thousand are believed to have been in the building at the time of the fire. Another common element shared by these two fires, as well as by major aircraft carrier fires and the World Trade Center disaster, is that the dead greatly outnumbered the wounded. Four hundred and ninety-two died in the Cocoanut Grove, compared with only 181 hospitalized with injuries, reversing the usual pattern of three injured for every fatality. Boston was fortunate in having many service personnel in the city, including trained firefighters and rescue crews and equipment to respond to what became within fifteen minutes a four-alarm fire, eventually upgraded to five. Due to an automobile fire in the next block, firefighters were on the scene of the Grove at the time the fire started at about 10:15 P.M.; but because most of the exits were locked and those that were open were jammed with bodies, they had considerable difficulty getting into the building.

The city had completed a major civilian defense exercise only the previous weekend and thus was able to organize medical assistance quickly. This training and administrative infrastructure proved invaluable, as Robert Moulton observed in his discussion of the first hour after the fire:

> It was estimated that one Cocoanut Grove fire victim reached the Boston City Hospital every 11 seconds, which is a faster rate than casualties were taken to any hospital during London's worst air raids. The magnitude of medical attention necessitated by this tragedy is reflected by the fact that more blood plasma for burn shock was used in Boston on the first day of the disaster than was used in Hawaii after the Pearl Harbor raid.[103]

Unlike Chicago in 1903, Boston had a well-organized, knowledgeable, and politically independent medical examiner, Timothy Leary, whose staff—most of it women serving as volunteers—did not permit uncontrolled claiming of unexamined bodies; quickly produced good records that allowed identification of bodies with a minimum of viewing; and separated bodies so that when viewing was necessary, mourners did not have to walk down rows of hideously disfigured corpses to locate the one they sought, as they had done in Chicago, and were to do four years later after the Winecoff fire.

Cocoanut Grove, however, remains for several reasons one of the great unsolved mysteries of U.S. fire history: first, because the cause of the fire is unknown; and second, because of the speed at which the fire traveled—nearly two hundred feet up a flight of stairs and across three large rooms, the better part of an entire city block, in something between seven and twelve minutes, leaving no chance for those in its path to flee. The third conundrum of this fire was why its smoke was so lethal. Very few died of burns, and some of those who died of smoke inhalation seemed uninjured after escaping the building, until they collapsed suddenly and died. Doctors at the hospitals reported finding few broken bones at autopsy, indicating that there was little panicked trampling or struggle. Even the two hundred or so bodies in the eight-foot-high pile behind the jammed revolving front door appeared to have been overcome as the fire passed over them rather than trampled or crushed.

The syndrome exhibited by the dead and injured seemed more consistent with exposure to phosgene than to ordinary combustion gases; but no possible source of this lung irritant was ever identified, although attempts were made to trace it to the leatherette wall coverings. In all, nine theories of the Cocoanut Grove fire have been published to date, including one involving an elaborate computer model of the fire; but each one, like the phosgene theory, has raised at least one unanswerable objection.

However the fire may have come about, it inspired another wave of closings for reinspection—more than a thousand in Boston alone—and revisions of building code both to include restaurants, bars, and nightclubs in the category of public-assembly occupancies and to prohibit flammable decorations and interior finishes in these structures. In communities that already had such provisions, there was a flurry (as always, temporary) of efforts to enforce them and to compel owners of precode structures to bring their public-assembly occupancies up to current code standards. Most of these improvements had to wait until after the war, however, when fire-resistive wall, floor, and ceiling coverings, most of them asbestos-containing, became more readily available for civilian uses, and there was sufficient labor to install them.

The second large multiple-fatality fire in the United States during World War II was the notorious horror of the Ringling Brothers fire of July 6, 1944, in which 168 persons were killed and 487 injured, about half of them children. As in the case of the Cocoanut Grove, the initial cause was unknown; but a tent sidewall was somehow ignited about five feet up from the tanbark, and the fabric, soaked in paraffin and gasoline to waterproof it, spread the flame quickly upward and outward. Because of wartime restrictions, the fire retardant with which the tent was usually treated was not available.[104] Even without the waterproofing, cotton has a flame-spread ratio of about 1,600; in other words, it spreads fire about ten times faster than hardwood does. Within minutes the tent's main poles were falling and the roof collapsing on the audience, which was frantically attempting to get down from the wooden bleachers, past the metal railings, away from the burning sidewalls, and over the projecting obstruction of the metal tunnel through which the circus' big cats entered and left the ring.[105] It was many years before Ringling paid off the debt created by its innovative installment-type settlement of the mass tort associated with the 1944 fire.

One of the many lessons of this fire was that no circus could ultimately afford to entertain a large audience in a structure as combustible as a tent; it was the beginning of the end for the big top as a circus trademark. After the war, circuses began holding performances in existing structures such as coliseums and arenas under local responsibility for enforcement of codes requiring fire-resistive materials in public-assembly occupancies.

Death on the Road: North American Hotel Fires, 1938–1946

For reasons I have explained, the decade of the forties was an exceptionally lethal era for public-assembly occupancies, including hotels.[106] These businesses had always had an ugly reputation for fires, ranking second after theaters, for example, in high fire-insurance premiums in London in the late nineteenth century. Even when inebriated guests no longer staggered up to their bedrooms with candles or fluid-fuel lamps, smokers still fell asleep in beds with burning brands in their hands, staff and guests made inappropriate uses of heating and lighting equipment, and hotel kitchens, like kitchens everywhere, were a perpetual source of fire danger. It was not unusual to try to save money, in the early to mid-twentieth century, by bringing in unlicensed and unqualified artisans to install electrical wiring in hotels, restaurants, and other small businesses, a practice that all too often proved to be a fatally false economy. In the early 1930s, arson was a frequent cause of business fires: owners who could neither make a profit nor sell their property opted to liquidate their insured holdings with a can of gasoline and a box of matches.

Prewar economic recovery in the American hospitality industry got off to a bad start in 1938 with a fire in the unfortunately named Terminal Hotel in Atlanta, a handsome five-story brick structure near the railway station, in which thirty-five of sixty-five persons in the building at the time lost their lives, many of them by jumping from windows, after a kitchen basement explosion ignited the combustible interior and roof. In April of that year, the Hotel Plaza in Jersey City also burned, killing four persons. The Queen Hotel in Halifax, Nova Scotia, was set ablaze by an improperly insulated furnace in 1939. There were no fire doors or other fire-resistive divisions of the wooden building, which even had wooden fire escapes. Twenty-eight died and nineteen were injured within ten minutes after discovery of the fire. In early 1940 the Marlborough Hotel in Minneapolis, another structure with combustible interior finishes, no fire divisions, and a wooden roof, burned with the loss of nineteen lives and twenty-eight injured.[107]

Once the war began, the high-occupancy pressure on lodging increased the risks. The Knights of Columbus hostel in St. John's, Newfoundland, for example, in late 1942, two weeks after the Cocoanut Grove fire, was in use as a temporary billet for members of the RAF and some U.S. service members. During a live radio broadcast from the hostel, a fire broke out in the all-wood structure, in which most of the doors had been locked and bolted from the outside. Of about five hundred persons thought to have been in the building at the time, ninety-nine lost their lives, and more than a hundred were injured. Arson was suspected but never proved.[108] On September 7, 1943, a fire at the Gulf Hotel in Houston—the type of accommodation colloquially known as a flophouse—took fifty-four lives when what was described as "a small fire of short duration" trapped men bunked down in rows of wooden cubicles in a brick and frame structure with inadequate exits, no exit signs, and no emergency lighting. There had been no building code at all in Houston at the time the structure was built. In Richmond, Virginia, another hotel burned in the spring of 1944, killing six after the combustible interior finish of the 1907 structure caught fire.[109]

The hospitality carnage continued after the war. During the night of June 5–6, 1946, a fire at the LaSalle Hotel in fireprone Chicago killed sixty-one of about a thousand guests in the building. A fire of unknown (probably electrical) origin started behind a seat cushion in the ground-floor lounge just after midnight and, with plenty of draft supplied by a ventilator shaft without firestops that ran up twenty-two floors next to the elevator, spread rapidly into the combustible suspended ceiling tiles and walnut-veneer paneling that were the hotel's decorative hallmark. This paneling was later tested and found to propagate flame at five times the speed of red oak. The hotel's stairs were fire resistive but were soon impassable due to smoke from the ventilator shaft and the unprotected doorway from the lounge; their uninsulated iron risers eventually failed from the heat. A guest discovered the fire about 12:20 A.M.; but because of hotel rules regarding staff authorization to call the fire department, firefighters did not receive the alarm until 12:35, an eternity later in firefighting terms and literally so for many of the guests trapped in inside rooms, below which flames and smoke filled the light well that was their only escape route. A brave and dedicated hotel operator asphyxiated at her post trying to waken and warn guests to evacuate the building. Many of these guests survived the fire, including some who used sheet ropes to descend to safety, a strategy that was to be less successful at the Winecoff a few years later. James K. McElroy, who described the fire for the *NFPA Quarterly*, clearly felt that Americans were not getting the message about the importance of fire-resistive construction and safety systems:

> Those who lost their lives in this hotel tragedy have joined the victims of past holocausts—the Triangle Shirt Waist fire, the Cocoanut Grove fire, the Hartford circus fire and others—who died that fire prevention and fire protection measures necessary to the safeguarding of the public be

inescapably drawn to the attention of architects and interior decorators, building officials, fire engineers, hotel managements and the traveling public.

As McElroy later observed, "just three days and four minutes following the holocaust at the LaSalle Hotel in Chicago," on June 9, nineteen more persons were killed in yet another hotel fire, this one in the Canfield in Dubuque, Iowa. Here, too, the combustible fiberboard dropped ceiling, added as part of an effort to modernize the appearance of the hotel, spread the fire through the building faster than the guests could get out of it and faster than the firefighters could get to the scene. Burns and asphyxiation on the upper floors were the principal causes of death.[110]

But it was the Winecoff Hotel fire in Atlanta on the night of December 7, 1946, that was the wakeup call for hotel occupancies in the United States.[111] One hundred and nineteen persons perished in an "absolutely fireproof" building, a designation that should by that time have already been considered a cruel joke: the Iroquois had been so designated in 1903 and the LaSalle as well, which had been built in 1909 and burned the summer before the Winecoff fire. As in the cases of the Cocoanut Grove and the Knights of Columbus fire in St. John's, U.S. service members and ex-military were among the dead. Winecoff victims included the veterans of fifty-three missions in a B-25 bomber, another of sixty-five bombing missions in a B-26, the Battle of the Bulge, Guadalcanal, and the sinking of the *Yorktown* at Midway. Among the dead were a marine who had come back from the Pacific theater, two army veterans, an air corps gunner, and a former Navy torpedo bomber pilot. Two of the victims were in Atlanta looking for postwar work on the night of the fire.

The Winecoff Hotel on Peachtree Street in downtown Atlanta was a fifteen-story structure, 155 feet tall, in a city with no fire ladders that would reach beyond the eighth floor—eighty-five feet. It was an "absolutely fireproof" building with a single stairwell wrapped around the elevator shaft, finished with burlap wainscoting, wooden chair rails, picture molding, carpeting, multiple layers of wallpaper, no landings, and no fire doors. The rooms were full of combustibles: wood, fabrics, bedding, furniture, and interior finishes. There were no fire escapes because the building was precode, built in 1913, when Atlanta's building code required only larger hotels to provide them. Atlanta did not require fire doors until 1924, so the Winecoff did not have them. Because of Depression budget cuts and continued postwar shortages, the Atlanta fire department was short of ladder equipment on the night of the fire and could not rescue guests on the upper floors as fast as the fire was pursuing them.

The top floor of the building was numbered sixteen because it would have been bad luck to have a thirteenth floor. At 3:30 on the morning of December 7, 1946, the fifth anniversary of Pearl Harbor, all of the Winecoff floors above the eighth were very bad luck indeed. As Sam Heys and Allen Goodwin point out, "more than eighty percent of the deaths occurred above the eighth floor, out of the reach of ladders." Of 280 persons registered at the Winecoff, 119 died and 64 were hospitalized, some of them with very serious injuries. Of the dead, thirty-two died after leaping or falling from upper-floor windows, some of them perishing when the sheets they had tied together to serve as ropes separated or were burned through when flashover fire burst through windows above them. Five of the victims were eight years old or younger, including a one-year-old baby. There was also a group of young people in the hotel, teenaged members of the Hi-Y, who were in the state capital to participate in a mock legislative session. Of forty Hi-Y members booked into the Winecoff, thirty were killed. These guests had been booked into large, inexpensive rooms for two or more persons each, with rollaway beds; many were at the back of the hotel where the ladder trucks could not maneuver in the narrow alley space, and some had fixed louvered shutters to provide visual privacy for rooms facing another building ten feet away. Above the fifth floor in the Winecoff, everyone in these shuttered rooms was killed.

As the Iroquois had done for theaters and Our Lady of the Angels was soon to do for schools, the Winecoff reignited a debate that had been smoldering for decades among fire-safety professionals and municipal government officials: could a municipality legally enforce current codes on structures built under an older code? City attorneys argued that this would be an unconstitutional ex post facto law; fire-safety experts argued that any government has not only the right but the responsibility to protect its constituents from obviously mortal danger as part of its police power. After the Winecoff and the Cocoanut Grove, fire safety won the argument for retroactive enforcement of building codes in public-assembly occupancies in most urban jurisdictions. It was the last triple-digit-mortality hotel fire in United States history.

In 1948, the *NFPA Handbook of Fire Protection* effectively synthesized the Winecoff Hotel experience with that of incendiary warfare in two trenchant paragraphs of a chapter on interior finishes and insulation:

> The behavior of combustible contents and interior finish materials under fire conditions was an important factor in the design and development of incendiary weapons during World War II in the United States and abroad. Hundreds of controlled tests were made in rooms simulating various loadings of contents, such as work benches, furniture, bric-a-brac, etc., with various types and locations for the source of ignition, under varied conditions of critical moisture content conditions [*sic*], species of timber, etc.
>
> These tests showed that there is a critical temperature, roughly defined during the war as a *flashover* point, at which all the combustible

> surfaces in a room burst into flame. The time interval between the ignition of an incendiary bomb and the time when *flashover* occurred was a valuable criterion in evaluating the effectiveness of various incendiary bombs. It has long been recognized that fires, first spreading slowly, will eventually reach the stage where all the combustible material in the fire area will flash into flame. No attempts had been made to measure such time intervals under controlled fire test conditions prior to the wartime research on incendiary bomb performance. While the nature of the phenomenon had not been critically studied or defined, its occurrence in fires was clearly recognized. The probability of *flashover* and the time interval depend on the character, amount and arrangement of combustible contents and also upon interior finish. Slow burning contents and interior finish will, in the event of fire, permit a longer safe period in which to call the fire department rescue forces and escape from the building than with quick-burning contents and highly combustible interior finish. In occupancies having moderate or low-hazard contents, the character and amount of interior finish materials are likely to be the determining factor in the hazard. In the Winecoff Hotel disaster (1946) the spread of fire from the third to the twelfth story, due to the unusual stairway arrangement, involved successive *flashovers* of the combustible wall finish of the corridors on each story, the fire increasing in severity as it progressed upward. [italics in the original]

The handbook went on to describe the veneered wood used in the LaSalle Hotel in Chicago, where "fire burned over the surface of the panelling five times as fast as over ordinary red oak lumber."[112] It was, in fact, the hotel fires of 1946 that inspired the fire-protection community to make the Steiner tunnel test of flame spread, which had been used in various forms at Underwriters' Laboratories since 1911 and was an unofficial standard since 1943 at the latest, become the national consensus standard as ASTM E84 and NFPA-255 in 1958.[113]

The Winecoff catastrophe made headlines all over the world and drew the attention of someone well placed both to appreciate the danger to the United States that fire then represented and to make sure that something got done about it. This person was President Harry S. Truman, who had in 1946 called a landmark conference in traffic and automobile safety that influenced federal, state, and local policy for decades thereafter. In 1947 he called a presidential conference on fire safety, attended by 2,000 delegates, at which committees drew up practical national policies and plans for reducing fire risk in American communities. In April 1947, a month before the conference met, a massive marine explosion in Texas City burned down much of the surrounding town, at a cost of at least 518 dead and more than 3,000 injured. This disaster, and those of the Winecoff and LaSalle hotels, were frequently mentioned in the proceedings of the conference.

As set forth in the program, the objectives of Truman's conferences were two:

1. Universal acceptance by the highest officials of the United States and municipalities of their direct responsibility for fire safety. The acceptance of the same principle by Federal executives charged with the responsibility for Federal properties is requisite.
2. Public support from all possible sources behind such officials in accomplishing the enactment and enforcement of adequate laws and ordinances for fire prevention and fire protection.[114]

Truman himself addressed the meeting at the opening ceremonies: "The Nation has been shocked by a long series of spectacular fires in the last few years—particularly in the last few months—which have resulted in such great loss of life and such widespread misery." He went on to mention the Texas City disaster and to add, "The great hotel fires of last year again showed that we cannot afford to entrust our citizens' lives to unsafe buildings." Two hundred and seventy persons had died in hotel fires in the United States in 1946.

The general chair of the meeting was Major General Philip B. Fleming, administrator of the Federal Works Agency; the executive director was A. Bruce Bielaski, a personage well known to many of the conferees as the assistant general manager of the National Board of Fire Underwriters. Attendees included a broad spectrum of persons interested in fire safety, including state and local government officials, insurance underwriters, schoolteachers and principals, firefighters, building owners and managers, environmental groups of every stripe, fire chiefs and marshals, architects, lumbermen and foresters, boys' and girls' organization leaders (including both Boy and Girl Scouts), retail and manufacturing associations, hospital administrators, attorneys, druggists, military veterans, granges, newspapers, the American Red Cross, Kiwanis and other service organizations, broadcast journalists, and bankers. No one could say that Truman's staff had not made every effort to be as inclusive as the protocols of the period would permit.

The committees appointed from this motley crew of dedicated fire-safety crusaders presented their recommendations to the president in the form of seven reports, an action plan, and a community guide: reports on firefighting services; fire prevention education; building construction, operation, and protection; laws and law enforcements; research; organized public support; and a final report delivered a year later on the progress of state and territorial programs to advance the cause of fire safety. In the intervening twelve months, only Wyoming had failed to act.[115] Both the Action Program and the Guide to Community Organization for Fire Safety were intended to help participants take home the lessons they had learned at the conference and mobilize their communities at the local and state levels.

High on the list of priorities was establishment of strict, up-to-date, and well-enforced building codes for fire-resistive building construction—the kinds of codes described in chapter 3 that required or recommended asbestos in many applications—in municipalities that had obsolete codes or, worse, no code at all. The Report of the Committee on Building Construction, Operation and Protection asserted, "In 1946, over 50 percent of the large-loss fires was outstandingly the result of inferior building construction." According to the records of the National Bureau of Standards compiled in 1945, of 3,322 municipalities, 1,342 had no building codes; another 612 had codes more than twenty years old, and 33 more had codes that were undated and thus were in all probability very dated indeed.[116] This state of affairs was quite properly regarded as a scandal in what was then the most industrialized nation in the world—indeed, at that time, the only nation with an intact industrial infrastructure.

President Truman, who had seen the destruction of Germany and Japan with incendiary warfare a few years before, knew that no homeland is secure in which 10,000 persons a year are dying in more than 830,000 fires. Two thousand fire-safety professionals, representing all of the forty-eight states, left his conference in 1947 with a mandate to do whatever it took to save lives in peacetime and prevent disaster in wartime by making the American built environment less susceptible to fire.

5

Schools, Homes, and Workplaces

Fire Prevention in the Postwar Built Environment

Asbestos emerged from the battles and firestorms of World War II with a global reputation for saving lives and property in a world haunted by images of burning buildings and bodies, a world in which whole cities had been obliterated by fire. In the United States, we were never again to feel safe behind our oceanic walls; we had carried fire to our enemies and knew that our enemies could—as they eventually did—carry fire to us as well. The engineers and underwriters who had drawn fire-vulnerability maps of Hamburg, Tokyo, and Dresden came home to communities that now looked to them like soft targets for incendiary bombing: frame-built residential districts with wooden-shingle roofs, schools with combustible interiors, grain elevators that needed only a spark to level the surrounding town.[1] These professionals set to work to make American farms, factories, and cities safer from fire by devising and promoting stricter building codes in the name of national defense. Most of these new and considerably expanded codes, both as they were enacted in American jurisdictions and as they were enforced by inspectors and insurance companies, included additional provisions in which asbestos was the material of choice.

American customers for asbestos products had to compete, however, with several other markets for a limited quantity of the mineral worldwide. Germany, Japan, and Belgium, which had learned the lessons of fire safety for national defense in the hardest possible way, were in the early 1950s among the largest customers for exported American asbestos fiber and products.[2] Nearly every European nation was rebuilding with more stringent fire-safety requirements and contributing to the demand for roofing and insulation. Britain was making use of its wartime experience to upgrade structural fire safety, conducting tests and experiments and publishing the results with recommendations for the use of asbestos for protection of structural members, roofing, floors, and interior finishes.[3] Even New Zealand, which had not been directly in the path of the war in the way, for example, London and Berlin had been, wrote new building codes and standards for fire-resistive construction that included asbestos.[4]

As automobile manufacturing reconverted, demand for brake linings and clutch facings soared. Furthermore, the United States and its allies never entirely demobilized from World War II and remained in what by the standards of the 1930s would have been considered an advanced state of military readiness for the remainder of the century and into the present. The program of stockpiling begun during World War II in the United States was continued in the Korean War and the cold war for the same reason it had begun in 1942: in some critical applications, the nation was dependent on asbestos supplies (those of southern African amosite, in particular) that had to cross an ocean.[5] Not until the late 1980s did the United States government finally sell off its asbestos stockpiles, well after the debate on the health issues associated with asbestos had risen above the level of internal medical debate and become a public policy question. This did not, as we have seen, prevent federal agencies (especially the Navy) from continuing to publish research that endorsed the use of asbestos in applications critical for fire safety.

American Workplaces on Fire, 1911–1964

Workplace safety in the United States made considerable progress after, predictably, a disastrous multiple-fatality fire in the early twentieth century. New York City's Triangle Shirtwaist fire of 1911 is justifiably famous for its gruesome circumstances and entirely gratuitous death toll; for the callous attitude of the business proprietors, who reportedly locked in their workers and could not be held legally liable when a major fire resulted in 146 deaths; and for the sweeping changes in building codes it inspired, particularly the development of an entirely new code category: building exits. The NFPA organized a Life Safety Committee in 1913 in response to this fire and in 1927 began publishing a separate handbook for exit codes, which has been revised as appropriate ever since.[6]

The furor over the unsafe conditions that the Triangle fire revealed in workplaces drove a nationwide movement to update, improve, and strengthen building codes, which, as we have seen, were central to the system of fire safety in which asbestos was an important component. Because laws everywhere were rewritten to make employers responsible for the safety of workplaces, the Triangle was the last single-structure workplace fire in the United States to claim lives in triple digits. This is the reason that David Von Drehle quite appropriately called it "the fire that changed America."[7] The full impact of enhanced safety consciousness in workplaces became apparent during and after the Second World War, when there was money for businesses to expand (sometimes in defiance of code, as we have seen). Moreover, in the case of military production, underwriters representing the federal government were on hand to inspect and enforce compliance. For this reason I have displaced the Triangle from the temporal sequence and elected to discuss it in the context of fire safety after World War II.

Like many of the other fires I have described here, the Triangle was a fire of combustible contents and interior finishes in a "fireproof" building, which remained standing and structurally sound, except for a failed fire escape and some of the windows, after fire gutted three of the ten floors. Like many other disasters, this one unfolded in a relatively short time—thirty minutes—but had an impact that was to alter public perceptions for at least a century.

The source of the fire was a wooden waste bin for cotton scraps that could hardly have been more fire-ready if it had been designed for the purpose. The source of ignition is not known for certain, although an unextinguished cigarette was thought the likeliest of several possibilities. Cutaways from piles of fabric laid out on tables for cutting into garment pieces were routinely swept into this bin, which was on the eighth floor of the Asch building, in which Triangle Shirtwaist occupied the three upper floors. Cotton fabric has a flame-spread ratio of 1,600 to 2,500, which is to say that it spreads flame nine to thirteen times faster than a pitch-pine board. The cutaways bins at the Triangle were cleared every few months by a recycler, then called a rag man, but not before hundreds of pounds of cotton scrap had accumulated in them. The loft spaces occupied by enterprises in the needle trades in New York City were not required to have sprinklers and typically had no fire divisions—no fire-resistive partitions and doors—in the large open rooms. The Triangle's floors were more than 9,000 square feet each, with no fire divisions, under twelve-foot ceilings ventilated with airshafts. Van Drehle suggests, with some evidence in his favor, that the proprietors of the Triangle, whose holdings were insured for considerably more than they were worth and who collected suspiciously often on fire claims in which only unsalable inventory was destroyed, did not really want the kind of usually effective fire-prevention system that was already in use in cotton mills, although not, as a rule, in garment factories.

In any case, once the paper patterns hanging over the cutting table on the eighth floor began burning, sending flaming pieces flying around the room, only about thirteen minutes elapsed before workers on the ninth floor, to whom the flames traveled before anyone from the eighth and tenth floors gave the alarm, were trapped between a stairwell on fire, a door that was locked, and a fire escape that had fallen from the outer wall, sending two dozen persons to their deaths. About fifty of the victims died on the pavement after jumping from windows or at the hospital soon thereafter, nineteen by throwing themselves on top of a descending elevator car, and the rest inside, from asphyxiation, burns, and trampling by their desperate co-workers. The fire started about

4:30 P.M. and was under control by 5:45 P.M.; the last body fell from the ninth floor at 4:57.

By the time the experience of this fire had been analyzed and assimilated by the fire-safety community, and responses were proposed in the form of stricter rules about exits and fire divisions that would allow workers to escape a fire, the textile depression of 1926 was already well under way, soon to be followed by the larger forces of the Great Depression. There was little new building and little money for enforcement of existing measures. During the war, as we have seen, only defense plants and dockyard facilities received much regulatory attention, although much was learned about how buildings and communities behaved under fire conditions.

After the war, however, fire engineers, local and state governments, and insurers set out to make American workplaces safer from fire and were remarkably successful in doing so. Not counting explosions, the United States did not have another triple-digit-mortality workplace fire until September 11, 2001, and this fire was not caused by carelessness or unsafe building design. Two trends, which had begun to emerge during the war, took on new momentum from fire-safety professionals with incendiary experience in the war, in-depth knowledge of major fires such as that of the Triangle and the Winecoff, and about a half-century of development of scientific methods for testing how materials and constructions held up during a fire. The first was a new emphasis on performance-based specifications for both building codes and industrial standards. In the case of fire resistance, as we have seen, the Steiner tunnel test, based on the performance of asbestos-cement as a zero-flame-spread material, became the standard for testing flame spread as ASTM E84. Other tests, such as those for certain kinds of electrical insulation, were clearly based on the performance of asbestos materials: the specifications correspond closely to scientific descriptions of the performance of asbestos-containing materials under test. New materials and products had to pass these tests, by this time performed routinely at the behest of manufacturers, usually by Underwriters' Laboratories, which published lists of approved materials for various constructions.

The second trend was toward much greater specificity in code requirements for certain types of construction, resulting in the inclusion in many model codes, especially those of insurance companies such as Factory Mutual, of detailed drawings of model construction techniques. Recommended or required materials in these constructions, including asbestos, were clearly indicated. Departure from these models was strongly discouraged by insurers and mortgage holders; mutual insurance companies would simply refuse the risk.

Many of the recommendations in these model codes were the results of very detailed and meticulous test programs. The use of asbestos as vent covers in grain elevators, for example, was advised after experiments with various materials were tried for this purpose and the results discussed by the NFPA's Committee on Grain Elevators. Dust in grain-storage facilities was both a fire and an explosion hazard; and in the case of an explosion, it was much better to vent the force of the blast than to allow it to blow out the building walls, possibly collapsing it onto occupants. Asbestos panels would not, of course, ignite and were light enough to open readily to the outside from the force of even a relatively minor explosion.[8]

High-rise office buildings, with steel structural members insulated against failure under fire conditions with asbestos, mineral wool, vermiculite, and other incombustible materials, were a relatively new type of construction after the war. The spray process for asbestos insulation had been invented in the 1930s; but between the sluggish economy of the thirties and the overwhelming needs of the military, especially the Navy, few civilian structures actually received this type of insulation until after the near-completion of the Navy's building program in 1944. Before April 1942, however, thirty-four federal structures in nineteen states and the District of Columbia had been sprayed with Limpet asbestos, not including ships.[9] Like all new construction methods, sprayed-on asbestos was rigorously tested for preventing the collapse of steel-frame buildings in fires in both the United States and abroad and found effective for this purpose. Fires in these buildings, such as that of a Montreal office building in 1963, in which there were no structural failures and no injuries in a thirty-six-story building with Limpet asbestos insulation on the structural steel, were the subject of considerable interest to fire-safety professionals.[10]

This type of insulation had the additional advantages of being fast and easy (although, it was discovered, hazardous to human lungs) to apply either at the steel plant or in the field and was joint-free after application. The most significant fire-safety problem with the sprayed-on asbestos was that contractors sometimes installed steel structural members sprayed at the steel mill, even if some of the insulation had come loose from the steel in transit, or removed it themselves to make repairs. Two floors of a fifty-story skyscraper at One New York Plaza in New York City were significantly damaged by a fire on the thirty-third floor in August 1970; investigators later discovered that the twisted steel beams had lost their insulation before being installed in the building and thus had no thermal protection. The building was nevertheless evacuated with only two fatalities resulting from a heat-activated elevator opening its doors on a fire floor. There was no structural collapse either above or below the fire floor. Testing of experimental steel structures in the 1970s unprotected by thermal insulation resulted in failure and structural collapse, a finding that was to resonate a few decades later when the collapse of the World Trade Towers was investigated.[11]

Dick and Jane Go to the Morgue: School Fires from Collinwood to Our Lady of the Angels

Few disasters devastate and demoralize a community more thoroughly than the loss of its children, and few arouse such passionate grief, indignation, and rage at whoever can be held responsible for the loss. The fire safety of American and Canadian schools was one of the first issues taken up by the NFPA after its formation in 1896, and the organization campaigned diligently for more than fifty years before local school systems began to take fire safety almost as seriously as they did their budget constraints. Members of the NFPA educated teachers and distributed classroom materials on fire safety; delivered lectures to parent-teacher organizations; and worked tirelessly to have all schools covered by appropriate building, life safety, and exit codes.[12] Their efforts had only limited success, however, until after the usual marker event in this history: a well-publicized major multiple-fatality fire. Between 1903 and 1939, the NFPA recorded sixty school fires involving loss of life, out of almost 1,000 total school fires, including 694 deaths that occurred in only eight of these fires.[13] There was no question that schools were in dire need of improved fire safety, and the data to support this position grew more appalling with every decade that passed.

As in the case of theater fires, there were actually two U.S. fires that bracket the period during which school fire safety rose nearly to the top of the collective national priority list for school facilities' security by 1960. Unlike the Richmond and Iroquois theater fires, however, only half a century separates the two most deadly American school fires; and only the second received a significant amount of publicity outside the immediate geographical area in which it occurred and, of course, the fire-safety community of its time.

The two events, however, had several elements in common: both involved failures to use fire-resistive materials in applications where building codes already required them; both structures were grandfathered into occupancy by having been built before the code was written; both fires spread at unbelievable speed from basement to second floor up combustible staircases and walls lacking firestops; both catastrophes involved significant loss of life from the lack of fire-protected exits opening outward; and most gruesome of all, both horrors occurred in full view of the community, including a crowd of parents who could see their children dying and could do nothing to save them.[14]

The first was the burning of Lakeview Elementary School in Collinwood, Ohio, now a suburb of Cleveland, on March 4, 1908—appropriately enough, Ash Wednesday. The school building was a brick structure with floors, stairways, roof, wainscoting, framing, doors, and other interior finishes of wood: in effect, a brick chimney full of fuel, with a combustible roof (see fig. 5.1). It had two stairways, one leading to the front door and the other leading to the rear exit.

FIGURE 5.1 The Collinwood, Ohio, elementary school after the disastrous fire of March 4, 1908. More than 170 people died in this building, the walls of which remained standing after the fire. An improperly insulated heating pipe, which, according to the investigating commission, should have been lagged with asbestos, was the probable source of ignition.

Photo courtesy of the Western Reserve Historical Society, Cleveland.

There was an attic in which classes were held, which, mercifully, had its own iron fire escape. The exact number of students in the building at the time of the fire is not known, but 366 were enrolled under nine teachers. Fire drills had been held regularly, with each class using its designated exit.

The village, which had a population of about 8,000 in 1908, had recently voted for annexation to Cleveland. Local politicians had been reluctant to invest in expensive fire equipment for its volunteer force, so the community was ill-prepared to respond to a major fire. When the alarm for Lakeview came in at 9:45 A.M., the fire horses were in use a mile from the fire station, pulling a road grader; replacements had to be hastily borrowed for the hose wagon. By the time the firefighters arrived at the scene twenty minutes after the alarm, the fire was already out of control. Somehow, the village's only fire ax was left behind at the station, and the ladders proved to be too short to reach the third floor. There was insufficient pressure in the leaky hoses to play water on the burning second floor. Most of the parents of the children in the school lived or worked nearby, and those who saw the smoke and heard the alarm came running to try to help by breaking the windows and helping the children climb out.

The back door was either partly locked or jammed, and few children got out through this exit. A pile of bodies five feet high was later found behind it. The fire escape worked exactly as designed: nearly all the students from the attic escaped.

Inside the school, a coal-fired furnace with a steam boiler had provided the heat. It was located immediately under the front entrance to the building, below the stairs, which were of Georgia pine; most of the furnace enclosure was wood. The steam pipes, some of which were reportedly lagged with asbestos, though at least one was not, passed through wooden floors; and either the uninsulated furnace or the unlagged pipe ignited the floor under the front hallway. Once the gong rang for fire evacuation, the classes began to file out in an order that normally took ten minutes to fully clear the building. On this occasion, ten minutes was five minutes longer than they had. Teachers tried to lead their classes to the doors they had been assigned, but there had been no instructions or practice with alternate exits if one was impassable. Some students and teachers on the ground floor of the school got out the front door by running through the flames, but after the first five minutes the wooden walls around the door were on fire, forcing the crowds of children and their teachers back on those still descending the narrow wooden stairs. A crowd of 250 children and teachers, in which two of the latter were crushed to death, fought to get out the windows while the fire climbed the stairs to the roof. The flashover of the crammed and panicked vestibule occurred in full view of parents trying to rescue their children from the windows.

The fire was pronounced out at 1:30 P.M., by which time 172 students and two teachers were dead; only about 64 of the schoolchildren in the building at the time escaped without injury. All of the student victims were between the ages of six and fifteen. The floors had collapsed into the basement, so the firefighters spent the rest of the afternoon hip-deep inside the water-filled brick foundation walls, trying to pull the charred remains of the village's children from the ruins with shovels, rakes, and pitchforks.

The best that can be said of this hideous episode is that it inspired a wave of fire-safety regulations and code revisions for school buildings that brought about some improvement in some communities. School fires, however, continued to be common occurrences in the United States. In Babbs Switch, near Hobart, Oklahoma, a frame schoolhouse caught fire from a candle on a Christmas tree and an overturned lamp, killing thirty-five. There were only two exits, and one of them was blocked.[15] The worst school disaster in American history was an explosion, not a fire: a gas leak from an unapproved fuel connection blew up a school in New London, Texas, on March 18, 1937, killing 294 children, teachers, and other staff members. In 1935, the *Handbook of Fire Protection* reported a fire record for schools that showed that 21 percent of the 869 school fires then recorded by the NFPA were started by defective, overheated, or inadequately insulated heating appliances and chimneys, as the Collinwood fire had been.[16] Handbooks and manuals of school fire safety, including that of the U.S. Department of the Interior's Bureau of Education in 1919, stressed the importance of asbestos insulation between heating appliances and wooden surfaces and of using fire-resistive construction materials, including asbestos products. Asbestos was also one of the recommended materials for school roofing.[17] Schools continued to be built without such protection, however, and with inadequate exits, especially rural schools outside the fire limits of municipalities.

The Cleveland School in Camden, South Carolina, for example, which burned in 1923, was built of pine lumber, lit with kerosene lamps hung from the ceiling, heated with wood, and provided with only one exit. Clara Elizabeth Hinson Woodson remembered it in 1997 as very similar to most schools in Kershaw County at the time, including one she attended later, after losing a brother, a sister, and a cousin in the fire that occurred in the Cleveland School on the evening of May 17, 1923, during a school pageant performance attended by all the students with their parents, teachers, and other family members. When one of the liquid-fuel lamps fell from the ceiling and broke on the wooden floor, the entire building was in flames in minutes. Seventy-six persons died, including, in some cases, entire families down to babies a year old: eight Davises, eleven Dixons, seven Hendrixes, and eight McLeods. The next school that Clara Hinson Woodson attended, constructed the same way, also burned to the ground; but because a state law passed after the Cleveland school fire required retrofitting of a second exit in these deadly wooden schools, everyone on that occasion evacuated safely.[18]

By the early 1950s, after efforts by fire-prevention activists, including parent-teacher organizations, the National Education Association, and the U.S. Office of Education of the Federal Security Agency, were beginning to bear fruit, the school fire rate was down to a "mere" seven to ten per day.[19] That this improvement was regarded as encouraging is evident from an April 1958 article in the *NFPA Quarterly*, which appeared a few months before the Our Lady of the Angels fire: "The 19 large loss school fires in 1957 caused property damage totaling $7,473,000 and resulted in 34 injuries, all to fire fighters. This experience is better than that of 1956 when 21 fires caused $9,033,500."[20] According to the Bureau of Vital Statistics' tables of mortality, about ten American children per day died in fires of one kind or another in 1950. This, clearly, was still too many, especially with the postwar baby boom just getting under way.

In March 1954, a single-story school annex in Cheektowaga, New York, burned, killing fifteen children when combustible ceiling tiles supported a flashover fire that cut off the only usable exit to a music room, which was fortunately the only room in use in the building at the time. Of thirty-four persons in the building, only two escaped unharmed because the only means of egress was through thick-paned windows that had to be broken to be used as exits.

At this period, dropped ceilings were beginning to be widely used in buildings for a variety of reasons: better acoustics, more modern appearance, reduction of heating and lighting costs, compatibility with the relatively new suspended fluorescent lighting, and concealment of unsightly building mechanicals such as ductwork and wiring. Unfortunately, as the 1946 LaSalle Hotel fire had clearly shown, ceiling panels could also conceal the spread of fire; fire-safety professionals have consistently inveighed against the dangers of concealed spaces in which a fire can flourish without detection. Dropped ceilings, especially those made of inexpensive and widely available combustible fiber, not only concealed the start of a fire but propagated flame at alarming speeds, as wartime and postwar tests confirmed. Fire-resistive, asbestos-containing ceiling panels were also widely available but cost significantly more than those made of wood, paper pulp, or other organic materials that spread flame at least as rapidly as wood. The school annex in Cheektowaga had been a temporary building erected by the federal government in 1941 as part of a defense expansion measure; the building was of frame construction with an asphalt shingle roof, and scarcely an ounce of then-scarce asbestos was used in the structure, which burned to the foundation. The two teachers believed that all children had been evacuated, until the west wall of the building fell, revealing the bodies of ten of them. The rest were found in the ruins of the hallway.[21]

Although the Our Lady of the Angels (OLA) school fire in Chicago on December 1, 1958, was not the most fatal of American school fires, it was to be the most influential, partly because it occurred at a time when both school fire safety and combustible interior finishes were well-known and frequently debated public policy issues. The Collinwood fire of 1908 was the subject of articles in the national press; but they did not have the impact and immediacy of the photographs and broadcast reports of the Chicago fire fifty years later, which included shocking images of firefighters carrying out one small, limp body after another from the smoking ruins. Moreover, the fire-safety community of 1908 was much smaller, less professionalized, and much less influential than that of 1958. The additional half-century of experience in post-fire analysis enabled fire professionals to be much more specific about how the fire had occurred and spread and what factors contributed to the loss of life. The relatively speedy extinction of the OLA fire also permitted examination of the building after the fire, a procedure that revealed much about the plight of those who had been trapped in it. The parents of the largest cohort of children ever to enter the American school system was at last ready to listen to experts who had for fifty years been telling them that their schools were lethally combustible firetraps.

The fire, believed by some to have been caused by arson, began, as the Collinwood fire did, in the school's basement under the north wing stairs to the first floor sometime before 2:25 P.M., the last hour of classes for the day. The first to actually see flames was probably the school's custodian, who had been at another property at the time the fire broke out; custodians were nearly always the staff member in schools most immediately responsible for fire safety since they attended to the most obvious hazard: the heating system.[22]

Although actually composed of three separate buildings (two were brick with wooden joists and one was of wooden-frame construction), the building was essentially all one fire area because it had only one class B fire door in a structure that housed, on December 1, somewhere between 1,200 and 1,300 persons.[23] The stairs under which the fire started were built of brick but finished with wooden lath and plaster on wooden furring strips. All the other stairways in the building, including those intended as exits in case of fire, were of unenclosed wood, wide open to smoke and flames and leading into corridors and classrooms finished with more wooden lath and plaster. Chester Babcock and Rexford Wilson, of the NFPA's Fire Record Department, reported, "Ceilings of all classrooms were finished with combustible cellulose fiber acoustical tile cemented directly to the plaster. It was also cemented to the ceilings of the first floor corridors of the north wing and annex." This tile, made of wheat straw, provided the fuel for flashover fires of the second-floor hallway and later of several of the classrooms, in some cases with students and teachers still in them. The two fire surveyors went on to say:

> Throughout this building there was wood interior trim in the form of doors, door frames, transom frames, mop and cork hook boards. Furniture was constructed principally of wood, and in the second story of the north wing there were pressed paperboard blackboards in the rooms and a large amount of children's clothing (it was below freezing outside) hanging from hooks along both sides of the main corridor.

All of this provided fuel for the fire that came up the stairways, cutting off exit to those in the back classrooms of the second floor even before the fire alarm was sounded at 2:42 P.M. By this time, students and teachers on the second floor were already at the windows, which were the only escape for most of them; only room 207 had access to a fire escape. Many jumped before fire ladders could be placed to rescue them; most of the jumpers survived the two-story drop. Some of the children were so small they had to be boosted to the windowsills. A crowd of shocked and terrified parents gathered around the school as the tragedy unfolded through the afternoon (see fig. 5.2).

Dense, choking black smoke and intense heat filled the corridors as combustion gases rose to the second-floor ceiling and crawl space above it, igniting the cellulose tile with a whoosh that survivors could hear in the classrooms behind their closed wooden doors (see fig. 5.3). After this, it was only five minutes—at 2:47 P.M.—before flames broke through the second-floor classroom transoms and flashed over five of the six rooms, killing everyone left inside and

FIGURE 5.2 Anxious parents watch as the bodies of schoolchildren are carried out of Our Lady of the Angels school in Chicago, December 1958. Combustible interior finishes, including ceiling tiles, made the school an inferno before most occupants even knew the building was on fire.

Photo courtesy of the *Chicago Tribune*. Reprinted by permission.

asphyxiating those in the sixth classroom, including several still at their desks. Thirty persons—twenty-nine students and a teacher—died in the flashover of room 210; the rest of the fifty-six fourth graders in the room escaped from the windows, some by jumping. The fire was brought under control by 3:45 P.M. and declared extinguished at 4:19, but by then ninety-three persons were dead, including ninety students and three teachers. Two more were to die in hospitals among the seventy-seven with serious injuries. One of them, thirteen-year-old William R. Edington, Jr., lay critically injured in St. Ann's Hospital for nine months before dying in August 1959. He was the ninety-fifth victim of the fire.

Parents, teachers, and school administrators all over the country looked at their school facilities with new eyes after the OLA fire and were dismayed by what they saw. There was a rush to enforce new codes on old school buildings,

FIGURE 5.3 Complete burnout of the second-floor hallway in Our Lady of the Angels school, December 1958. The ceiling tiles, of which few traces remain in this photograph, were made of wheat straw. Flames from the hallway flashed over through the transoms of two of the classrooms, incinerating some people and asphyxiating others exposed to the smoke.

Catholic World, 1958. Photo courtesy of The Catholic New World, Archdiocese of Chicago.

a controversial measure that had long been advocated by fire-safety professionals and opposed by most school districts. Set against the backdrop of grief and loss in Chicago, the legal and budgetary objections to retroactive code enforcement began to seem trivial and callous, politically damaging enough to provoke action on the part of local authorities. Regional, state, and national meetings of educational officials were convened, and codes of best fire-safety practice were adopted in the construction of new schools and the renovation of old ones.[24]

These codes, as we have seen, included the use of noncombustible interior finishes, especially for corridors and exits, including those that contained asbestos. If Our Lady of the Angels had had asbestos-containing ceiling panels, flooring, and wall finishes, there would almost certainly have been time for all the building occupants to reach the exits and little or no loss of life. The 1960 report of the National Academy of Sciences' Committee on Safety to Life from Fire in Elementary and Secondary Schools contained a discussion of flame spread, mentioning with approval those materials that scored at or near zero on ASTM E84: "non-combustible materials such as asbestos-cement board, plaster, concrete and ceramics." By 1964, recommendations for fire-resistive materials, including asbestos, were included in the National Council on Schoolhouse Construction's official guide for school planning.[25]

The role of combustible ceiling panels, so significant in the loss of life in the Chicago school fire, had drawn the attention of fire-safety professionals after the LaSalle in Chicago and other hotel fires in the 1940s and were also an issue in other types of high-occupancy and institutional fires, such as department stores and armories. Building inspectors were advised by fire engineer Robert Moulton in 1949 to examine fiber-based ceiling tiles carefully to be sure they were of "noncombustible fibres such as asbestos or glass fibre" in public assembly or institutional occupancies. Even as late as 1956, many such structures were not using appropriately fire-resistive materials in venues from which the occupants might be physically slow in evacuating, such as hospitals and nursing homes, or even restrained by locks and bars from escaping, such as prisons and jails.[26]

No Exit: Protecting the Incarcerated and Mobility-Impaired

When chemist and military workplace safety consultant Peter J. Hearst prepared his 1980 report for the U.S. Navy's Bureau of Medicine and Surgery on recommended materials for hospital corridors, he endorsed well-maintained asbestos-cement as the safest choice in this application, despite the known inhalation hazards of the material. Hearst was familiar with the ugly history of hospital fires and was making an explicit tradeoff between the acute risk of fire and the chronic, long-term risk of asbestos-related disease.[27] His concerns were justified: in the same year that his report was published, fourteen persons died in a boarding home in Ohio in which the ceilings and wall paneling were combustible fiberboard; two years later, thirty-four more were killed under similar circumstances in Illinois and New Jersey.[28]

As in the case of schools, fire-safety professionals argued for decades for fire-resistive construction and interior finishes in hospitals, nursing homes, and board-and-care facilities before these institutions acquired a shameful record of multiple-fatality fires that continued even longer than that of schools. Sprinkler systems for institutional fire protection were also endorsed by the National Board of Fire Underwriters and the NFPA, but safety experts were explicit in their assertions that neither the passive protection of fire-resistive construction nor active suppression systems could be relied on by themselves. Both were necessary, but neither was sufficient, especially not in venues such as hospitals, nursing homes, prisons, jails, and orphanages in which occupants would need assistance in evacuation.[29] Few such institutions could be evacuated in less than thirty minutes, even under the best of circumstances, and evacuation times of up to two hours were not unknown. Fires in combustible structures or those with combustible interior finishes are rarely this generous with time.

By the time Hearst wrote, the fire-safety situation for institutions had been greatly improved, although fatal hospital and nursing home fires continued to occur. Between 1915 and 1929, however, the NFPA's Department of Fire Record collected reports of 184 such fires, 33 of them fatal, accounting for the loss of 273 lives, or about 8.3 deaths per fatal occurrence. Of these deaths, 130 were children, 53 of them babies killed in a single fire in the Grey Nunnery in Montreal in 1918. Another 37 girls between five and nine years of age were lost in the Hospice St. Charles in Quebec in 1927. In 1929, the year of the NFPA report, the National Board of Fire Underwriters reported that North American hospitals and nursing homes were burning at an average rate of more than one per day. Many of these institutions were located outside urban fire limits, unhampered by the constraints of municipal building codes, and few states had or could enforce codes before 1930. Combustible construction and interior finishes were nearly always implicated in institutional fires. The wooden roof shingles of a single maternity hospital, the Midwood Sanitarium in Brooklyn, New York, caught fire three times before 1928. Institutions that stored nitrocellulose X-ray film were especially at risk; even a small fire in a film storage room could result in asphyxiations all over the building as the toxic fumes spread through the heating and ventilation system, as they did at the Cleveland Clinic in 1929, where 125 persons lost their lives.[30]

Mental hospitals were nightmares in fires: the patients were not only locked in, but many could not be made to understand their danger and had to be physically subdued in order to be safely evacuated. A fire in the jerry-built wooden-frame structures of the Central Oklahoma State Hospital for the Insane killed thirty-eight inmates in 1918; and Manhattan State Hospital in New York City, at that time one of the largest such institutions in the world, lost twenty-two patients and three staff members in February 1923. Eighteen died by fire in December of the same year at the Illinois State Hospital for the Insane. When the polished wooden floors, stairways, and corridors of the Little Sisters of the Poor Home for the Aged in Pittsburgh, Pennsylvania, burned in October 1931, forty-eight persons were killed, most of them by smoke inhalation.[31]

In January 1945, a small hospital for babies in a wooden-frame structure in Auburn, Maine, caught fire and killed sixteen of its infant patients. In December of the same year, fire professionals saw once again the dangers to the other end of the age spectrum when the Niles Street Convalescent Hospital burned in Hartford, Connecticut, a city that became notorious for nursing home and hospital fires, despite being the fire-insurance capital of the United States. The fire started with the lights of a Christmas tree on December 24, 1945, killing seventeen patients and two attendants. Thirty persons were injured, including ten firefighters. The structure was brick, but all interior construction was wood, and the wiring had been cobbled together without the benefit of a professional electrician.[32]

Worse was yet to come. Fire-safety professionals were already learning in the late 1940s, mainly from hotel and other assembly fires, that the fashionable dropped ceilings that were then being widely used to modernize older structures were fire hazards unless made of incombustible materials, of which mineral tile—usually asbestos—was regarded as the top of the line as far as fire safety was concerned. It was also, unfortunately, more expensive than combustible fiberboard, from which it was visually indistinguishable from below; and cost-conscious builders, administrators, and renovators persisted in their use of these well-known fire hazards until well into the 1960s. On April 4, 1949, a fire in the St. Anthony Hospital in Effingham, Illinois, killed seventy-four of just over a hundred patients in the hospital, when fire in a wooden laundry chute connecting all three floors spread to combustible ceiling panels and open wooden stairways in a building with no fire divisions except one between the laundry and the rest of the basement. Despite the prompt arrival of the fire service, the building was almost entirely destroyed, including the hospital's records of what patients were in the building at the time of the fire, making later identification of bodies difficult. The bodies of eleven newborn babies in the maternity unit had to be identified by the metal bead necklaces placed on them at birth, as soldier's remains are identified by their dogtags.[33]

Combustible-tile dropped ceilings were implicated in the deaths of forty-one of the sixty-four women patients in the St. Elizabeth's Women's Psychopathic Building of Mercy Hospital in Davenport, Iowa, the following year. Twenty more perished in a ceiling-panel fire in Hillsboro, Missouri, in 1952, and another thirty-three in Littlefield's Nursing Home in Largo, Florida, in 1953, in which both ceilings and wall partitions were constructed of combustible fiberboard. An interior finish fire at the Hartford Hospital in 1962 took sixteen lives, including seven patients, four staff members, and five visitors. Fire investigators rated the flame spread of the ceiling panels at 180 and that of the non-asbestos linoleum wainscoting at 300.

As if the fires in the health-care facilities were not enough, combustible interior finishes were still being implicated in multiple-fatality hotel fires in the 1970s and 1980s. Clearly, fire-safety professionals had excellent reasons to endorse the use of incombustible materials, including those that contained zero-flame-spread asbestos, in the construction of hospitals, nursing homes, and board-and-care facilities. While fires still occur in these structures, far fewer of them are fatal since the majority have adopted the system of fire safety developed in the first half of the twentieth century.[34]

Fires in prisons and jails obviously created the same kinds of problems for evacuation as those in mental institutions, with the added danger posed by having production facilities, such as those for license plates, with all their industrial-occupancy hazards, housed in the prison building. As in the case of schools and hospitals, fire-safety professionals began pleading as early as 1913 for fire-resistive construction of correctional facilities after thirty-five African American prisoners in Jacksonville, Mississippi, died in a wooden cage on the second floor of a building in which hay, corn, and molasses were stored on the lower floor. "What can the poor convict expect of our humanity while the factory workers and even the school children are not properly cared for?" an editorial in the *NFPA Quarterly* asked in the wake of this fire, only two years after Triangle Shirtwaist.[35] The Ohio Penitentiary fire of 1930, described in chapter 1, in which 320 inmates perished, shocked prison officials into much greater care in the construction of prisons for fire safety, but many jails remained firetraps.

During the period in which asbestos use was being greatly reduced in construction after 1970, serious fire hazards from newer materials began to appear. A foam-plastic material on the walls of padded isolation cells in the Biloxi, Mississippi, jail in 1983 spread flame, smoke, and heat that killed twenty-nine when officers could not open the cells fast enough to release them.[36] Toxic fumes from burning interior finishes were implicated in fires in more luxurious accommodations than the Biloxi jail: in 1980, eighty-four died in the MGM Grand Hotel. The fire was started when an aluminum electrical conduit melted and ignited nearby plywood. Twenty years before, conduits of this type had been routinely insulated with asbestos, which would not, of course, have melted. Combustible interior finishes had spread flame in the Beverly Hills Supper Club fire in May 1977, in which 167 lost their lives, foreshadowing the tragedy of the Station nightclub fire of 2003.[37]

Living under a Cloud: Fire-Resistive Communities in the Cold War

In 1950, *Mill and Factory* published a lavishly illustrated article called "What Is Asbestos Good For?" which listed scores of industrial applications for the mineral. The editors asserted, "Asbestos helps man and industry defy these public enemies: fire—weather—heat—acids."[38] The article was typical of postwar engineering literature about asbestos, for which new uses continued to be found, including valves, ductwork, and pipes of all kinds but especially those that were subject to corrosion from acids or sewage. Asbestos manufacturing flourished, more than tripling its work force between 1930 and 1970. The *Occupational Outlook Handbook*, published by the U.S. Bureau of Labor Statistics, asserted in 1951, "The outlook for asbestos workers [in insulation and construction] is good. This trade is employed extensively in some kinds of defense construction and is important in peacetime." Wages, which were "in accordance with collective bargaining agreements," ranged from $2.05 per hour in the south to $2.75 in New York City, about the same as for roofers. No unusual occupational hazards were mentioned in this or the subsequent (1957) edition of the handbook, in which government analysts predicted substantial growth in employment through the 1960s "as a result of the anticipated sharp rise in the volume of construction of commercial and industrial buildings" and the increased use of industrial pipe and ductwork in the petroleum and chemical industries, refrigeration equipment, and air conditioning.[39]

At this period, asbestos was thought to be good for practically everything from dentistry to rockets. *Workbasket* magazine and the Spool Cotton Company, makers of J.&P. Coats yarn and thread, published patterns for colorful crocheted-lace "Hot Plate Mat Covers" to liven up the appearance of the 8½-inch round gray asbestos pads that had been available for kitchen and stove use since the end of the nineteenth century. Typically, these covers were crocheted to match the ornate potholders popular in the 1950s.[40] According to the *Science News Letter*, it was possible in 1948 to buy dishtowels of an asbestos-cotton blend that would double as extinguishing blankets for kitchen fires. Schools used asbestos-containing materials for papier-mâché and modeling clay, a practice that experienced its fifteen minutes of infamy in 1983.[41]

Mining, of course, benefited from the great expansion of markets for asbestos after 1945 and the apparently insatiable worldwide demand for the mineral. In the 1930s, a nine-tiered system of grading asbestos had been developed by the Quebec Asbestos Producers' Association, in which group 1 was crude asbestos with a staple (fiber) length of ¾ inch or longer and group 9 was asbestos-bearing gravel and stone. Before the 1920s, there had been few uses for any but the top two grades, then called "Crude No. 1 and 2," and thus no need to classify the remaining products of mining with staple lengths under ⅜ inch; but as new uses, primarily in thermal and electrical insulation, began to be developed for even the "floaters" (sometimes called "refuse" or "shorts"), both sellers and buyers needed a system for determining quality and price.[42]

This trend had three effects on asbestos mining. First, it made previously marginal deposits of short fibers economically attractive, with the result that old deposits and their tailings could be reworked into the shorter grades of fiber; and new mines opened to exploit deposits that had not previously been considered worth the trouble. New mines were opened overseas in Morocco, Egypt, China, and Australia and in Maine and California in the United States.[43] Even Alaskan "mountain leather," or paligorskite, a dark and usually soggy asbestiform mineral resembling untanned hide when dry and previously thought unworkable, was the subject of experiments in the making of filters and acoustical and shock insulation. Alaskan tremolite had reportedly been used by the U.S. government during the war as a filtering agent for blood plasma.[44]

Second, demand encouraged the exploration of new uses for the material since in the lower grades it was no longer as expensive relative to other materials as it had been in the 1920s. It was already in use as a filler for plastics, including so-called "Ebonized asbestos," in electrical insulation applications such as switchboards in the 1930s; and additional uses for asbestos combined with rubber or phenolic compounds (plastics) were added almost every year during the late 1940s, 1950s, and 1960s. Ebonized asbestos was approved by the U.S. Navy Bureau of Engineering in the 1930s and was widely used in industrial electrical installations as well. The material reportedly would withstand temperatures as high as 2,800 degrees Fahrenheit but could be worked with carpenter's tools, unlike its predecessors in the same applications, marble and mica.[45] It also weighed considerably less than either.

Some of the new uses for the mineral were relatively glamorous—literally space age, such as the phenolic-impregnated asbestos flame shield adopted by the U.S. Army for the Pershing missile and the simpler asbestos insulation used in backyard rocketry by teenagers. The product manager of Raybestos-Manhattan described the new phenolics with evident patriotic pride in *Plastics World* in 1958:

> A special family of asbestos-reinforced plastics materials has been developed here at Raybestos-Manhattan for specific use in missilry [*sic*]. These materials have gone into missile fins and shrouds, rocket launcher bodies and tubes, rocket throats and cones, silver traps for propellant barriers and rocket tailcap insulators, to mention a few.
>
> Titan, Polaris, Sidewinder and Terrier missiles are using laminated or molded asbestos as heat insulation shields, tubing and heat diverters. The Navy's Vanguard is equipped with an asbestos-phenolic nose cone intended to streamline the rocket and to insulated the enclosed satellite

> from friction-generated heat in high-speed upward flight as well as to protect the rocket itself from the ram effect of aerodynamic loading.
>
> These parts for missiles feature high-strength to small-weight ratio (an important fact in rocketry), an extremely high modulus of elasticity at low and elevated temperatures, resistance to blistering, delamination and erosion when subjected to high temperatures and thermal shock.

In 1964, the year that the New York Academy of Sciences published its volume of papers on asbestos, two especially flashy new uses for asbestos were announced. One was "the extension skirt for the second-stage engine of [the] Gemini 4 space rocket [which] was made of asbestos felt impregnated with a specially developed phenolic resin;" the other was powdered asbestos "used as the reinforcing agent in the joining of the various reinforced plastic parts" of the Chevrolet Corvette automobile body.[46] Other uses of the period were more pedestrian: heat-resistant ball valves for oil and gas pipes, asbestos-nitrile rubber gaskets used in off-road equipment and compressors, asbestos plastic bearing materials, and fire-curtain walls for railroad engine roundhouses.[47]

Third, the marketability of floaters and refuse encouraged mining and manufacturing companies to install more effective dust-removal systems. Efforts to control dust for health reasons in asbestos factories and mills had begun during the 1920s; and after publication of a U.S. Public Health Service standard of 5 million particles per cubic foot of air in 1935, the American Conference of Governmental Industrial Hygienists and many states adopted this figure by 1950 as one below which "new cases of asbestosis would not appear," as the health service expressed it.[48] Predictably, industrial firms responded grudgingly to this standard, much as coal mines did when measures to reduce black lung were introduced and as cotton mills were to do later when required to reduce levels of the airborne cotton trash that causes byssinosis.[49] Between the reluctance of manufacturers to spend money on dust-control equipment in a period of slack business, the parlous condition of state labor department inspection budgets during the Depression, and concerns by unions and others dependent on the income stream from what manufacturing jobs had survived the downturn of 1929–39, dust-control standards in many U.S. industries were not often vigorously enforced in the decade immediately preceding World War II.

Dust-control equipment originally installed to protect the health of workers, however, began to seem more economically reasonable when it collected a product that could be sold at a profit; and measures were taken in most mining and manufacturing operations that handled raw asbestos to recover as much of the dust as was possible with the technology of the period. Even good dust-collection technology still left a great deal of dust on floors, ceilings, walls, machinery, and other surfaces, a problem that had still not been solved by the mid-1980s. Nevertheless, companies that could afford the outlay during the postwar boom, such as Johns-Manville, invested in the most sophisticated dust-control mechanisms available, especially for new plants.[50]

Many American industries, of course, prospered after World War II because in 1945 the United States was the only major western power with an intact industrial base. Five additional factors contributed so significantly to the prosperity of asbestos manufacturing that the U.S. Bureau of Mines' annual *Minerals Yearbook* reported every year from 1945 through 1952 that world demand for asbestos and its products exceeded the supply.[51]

First, there was the fact that the United States did not fully demobilize from World War II and became involved in the Korean conflict within five years of the end of hostilities with Japan. Experience during the world war had impressed on policymakers the importance of having significant stockpiles in the United States of raw fiber of the critical-materials type, particularly amosite, and of maintaining productive capacity for such military necessities as clutch facings and the high-pressure boiler packings used by the Navy. Stockpiling efforts were complicated by competition from the private market, as Oliver Bowles and F. M. Barsigian observed in 1950: "Industrial demand, much of it for defense orders, was so high that it was difficult to obtain material for the National Stockpile. Some progress has been made in developing substitutes."[52] Demand for all grades and types of asbestos created economic pressure to develop and improve substitutes such as fiberglass that could relieve shortages in civilian applications such as theater safety curtains and electrical insulation. In 1952, *Chemical and Engineering News* noted, "There is no known substitute for it [asbestos] in friction materials (for example, brake and clutch linings for the transportation industry); there is no known substitute for asbestos used in steam packings," and went on to add, "Glass fibers have been found to be suitable for low-temperature insulation and for electrical wire insulation. The vitreous silica types withstand higher temperatures than the soda lime-silica type, but replacement of asbestos where elasticity or flexibility is important is doubtful."[53] Even with substitutes in use in some applications and ongoing federal stockpiling efforts, Robert Keating reported to the Committee on Undersea Warfare of the National Academy of Sciences and the National Research Council in 1962: "Asbestos reserves in the United States are incapable of meeting either current or projected demands."[54] In order to build up the stockpile, the Commodity Credit Corporation "let contracts under the barter program in 1961 for 5,000 tons of amosite and 10,000 tons of crocidolite to be delivered by 1963 in exchange for surplus agricultural commodities," mainly with developing nations. This program continued through 1967.[55]

Second, there was a critical worldwide shortage of building materials and intermediate industrial goods, such as brake linings, gaskets, ductwork, and pipe in countries in Europe and Asia whose urban and industrial infrastructures had been almost completely destroyed during the war years. Developing nations wanted to start up their own industrial operations, and for this they needed electrical generators, boilers and their insulation, cable, and hundreds of other items in which asbestos was used. The United States was the leading supplier of the free world, the Soviet Union the leading supplier of the smaller nations under its economic influence.[56] Exports of manufactured asbestos goods from the United States continued to increase from their already high 1940s levels into the 1970s.

Third, returning veterans coming home to jobs in the expanding economy used train, bus, and truck transport and, within a few years, were ready to buy automobiles, all of which employed asbestos as insulation or friction material or both. Fourth, as this generation married and generated the baby boom, the demand for housing and schools kept the demand for fire-resistive building materials high.

Fifth, there was the new threat of nuclear war. Fire-safety professionals from the Allied nations had gone to war between 1939 and 1945 with respect for fire well planted in their imaginations; they returned from it terrified about the vulnerability of their own nations to the great scourge of incendiary warfare. Moreover, some of them had seen in person, and others in photographs and newsreels, the devastation of structures and destruction of life in Hiroshima and Nagasaki and realized that dishing out this kind of warfare meant learning to take it. The exigencies of a possible nuclear war involving the continental United States became a priority element in fire-safety planning, including new attention to building construction, urban development, and civilian defense.

During the war, there had been significant growth in the standardization of specifications for industrial materials and increased emphasis on methods of test so that military inspectors could be certain that they were receiving the quality of goods they had specified and that the taxpayers had paid for. Most of this standardization was immediately recognized as highly beneficial to industry because it made the merchantable qualities of products measurable and verifiable by objective third parties such as Underwriters' Laboratories and Pittsburgh Testing, protecting both buyer and seller. UL had been inspecting, testing, and certifying the safety of a wide variety of products since the early twentieth century, but their standards and methods of test were not always subject to consensus review by fellow professionals. The tunnel test for flame spread, for example, although in use at UL since the second decade of the twentieth century, did not become ASTM E84 until more than one Steiner tunnel was available for testing purposes. The new postwar edition of the NFPA's *Life Safety Code* reported flame-spread classes for nearly all commonly used building materials based on how they compared with the asbestos-cement used as the baseline in ASTM E84.[57]

The National Bureau of Standards (NBS) had its own product testing facilities and assisted UL, ASTM, and other professional organizations in both testing methods and development of standards. Both UL and the National Bureau of Standards, for example, tested and rated asbestos-containing fire-resistive doors as well as other fire-safety products.[58] As asbestos-cement panels, siding, and shingles gained ground in the postwar market, selling at "three times their prewar rate," NBS issued two sets of test results for these products in its *Building Materials and Structures Report* series in 1951: one by Cyrus Fishburn on the properties of asbestos-cement siding, the other by Nolan D. Mitchell on fire testing of walls and partitions faced with asbestos-cement. (Mitchell had performed the NBS tests of asbestos theater curtains a decade and a half earlier.)[59] Architect Paul Schweikher was commissioned by the Asbestos Cement Products Association in 1954 to compile a portfolio of asbestos-cement siding design ideas, in which he remarked that the material was "adaptable to rapid assembly and offering durability, low maintenance, harmony in color and uniformity in texture" in addition to the fire resistance that was a major issue in the postwar housing market. By 1968, half the world's asbestos was used in asbestos-cement products, including pipe and ductwork, and by 1969 it had climbed to 70 percent.[60]

ASTM's *Book of Standards* became first a multivolume set and then a triennial compendium resembling—and functioning as—an encyclopedia of product standards. Methods and standards of test that had been in development since the 1920s passed swiftly through the professional consensus review process during the war and were revised regularly thereafter, accompanied by much discussion in the engineering literature. Even products that had been in the marketplace for decades were tested and their specifications standardized; in the 1960s, ASTM published standards for asbestos-cement fiberboard ceiling panels, siding, clapboard, shingles, flat and corrugated sheet asbestos, and various kinds of pipe, spelling out how much organic fiber was permissible in each.[61]

It was not only engineers who valued the application of standards and scientific methods of performance testing. Even consumers wanted their products tested and rated. The postwar era saw the rise of consumer rating magazines, which published the results of performance tests in a format that could be easily understood by a nonprofessional. The *Consumer's Research Bulletin*, for example, rated six roof coating materials in 1952, all of which conformed at least nominally to federal specification SS-R-451, which required that such coatings contain a minimum of 5 percent and a maximum of 25 percent asbestos fiber. Two were recommended: Flintkote Static Asphalt (Fibrated), and Johns-Manville Asbestos Fibrous Roof Coating. Another product, a Japanese-made fire pouch of asbestos, did not fare as well when *Consumer Bulletin* evaluated it in 1968; "complete destruction" of the enclosed papers "took place in a very few minutes."[62]

In the engineering community, ratings for fire, electrical, and thermal resistance were applied to all materials, including asbestos, and much finer distinctions made between types and qualities of the same material. One example was the testing of chrysotile used for electrical insulation, which had, before and during the war, been evaluated by what was called the Navy test—military specification MIL-I-3053A. In the late 1950s there was a minor controversy over this standard, with some engineers calling for its replacement with what they considered a more accurate method, ASTM D1118-57, because the latter allowed measurement of the size of the magnetite particles, a factor in electrical conductivity, rather than just the total magnetite content by weight as measured by the Navy test.

The *Fire Protection Handbook*, the journal *Electrical World*, and the *National Electrical Code* published tables of what kinds of insulation were appropriate for particular applications; of twenty-six insulating materials listed in article 310 of the *National Electrical Code* in 1947, nine included asbestos. For applications in temperatures over 105 degrees Celsius, such as the arc proofing of manhole cables, asbestos was the preferred material.[63] Mechanical and civil engineers also made careful studies of the comparative advantages of various insulating materials; asbestos topped Robert J. Fabian's table of "How Insulations Compare," published in *Materials in Design Engineering* in 1958. When asbestos lagging cloth became available with a water-soluble impregnated adhesive in 1967, reducing the risk of chemical burns to workers' hands, the new product was tested and approved by the Coast Guard and in military specifications for fire-resistive materials.[64]

Similar professional discussions among engineers were going on about other traditional and new uses for asbestos. As we have seen, proximity clothing was significantly improved during World War II; and in 1956, aluminized asbestos fabric was introduced by the Minnesota Mining & Manufacturing Company (3M). A dozen companies immediately began making proximity suits and other hot-work gear from the material, which was much lighter than 100 percent asbestos and allowed wearers to walk through temperatures of up to 3,000 degrees Fahrenheit and to work for a few minutes at a time at temperatures of 1,400 degrees Fahrenheit. The new material passed all tests with flying colors. In 1963, Bethlehem Steel began experimenting with asbestos-polyester protective clothing for molten metal handlers, who had previously worn asbestos-cotton suits. This experiment also proved a success, with the new garments lasting half again as long as the older fiber blend. Proximity clothing proved to be a difficult problem when substitutes for asbestos began to be sought in the 1970s and 1980s.[65]

Asbestos had been in use for ductwork before and during the war; but like other asbestos products for construction, production and sales grew significantly after 1945. Experiences with wooden ductwork during the war had not been happy, and no form of combustible ductwork was code-compliant. Asbestos ductwork had been tested and evaluated in the usual way in 1938, with results of testing for friction losses, heat transmission characteristics, and acoustic qualities published in 1939. After two more decades of experiments with this type of ductwork, engineers from Johns-Manville reported in the *ASHRAE Journal* that because of the smoother inner surfaces and "unobstructed joints" of the mineral-based product, asbestos ducts reduced pressure drop by 30 percent. These ducts were never as widely used, however, as asbestos-cement panels and pipes, possibly because they were expensive, liable to breakage in handling, and could not be bent to shape when necessary, as sheet metal could.[66]

Asbestos-cement pipe made great market gains after the war both in the United States and abroad. It had been made and used in Europe since 1913 and in the United States since 1932; but because of market conditions during the Depression, only fifteen miles of the pipe were manufactured in that year. In 1947, *Scientific American* called it a "Pipe with a Future," and so it proved: by 1956, the United States was turning out fourteen thousand miles of asbestos-cement pipe annually. During the war, it had been substituted for strategic and critical metals, but it was soon found that asbestos-cement pipe worked well even if one used it on purpose. The price was not much lower than conventional materials, but the lower costs of installation and maintenance made it an attractive option for cost-conscious municipalities and manufacturing operations.[67]

Like the other applications I have described here, asbestos-cement pipe was extensively and rigorously tested by ASTM, UL, and other engineering organizations. It was sufficiently durable to allow at least one community in the 1960s to dig up and reuse for municipal water supply asbestos-cement pipes from the abandoned anti-aircraft unit that had protected Hanford Atomic Energy Plant in Washington State during the war. Unlike iron and steel, asbestos-cement pipe would not corrode, even in very acidic soils; did not rust or tuberculate; would not heat up and ignite surrounding combustibles; and was not electrically conductive, which meant, first, that it would not short-circuit nearby wiring and, second, that it was not susceptible to electrochemical corrosion. Unlike iron pipe, it did not oxidize and therefore did not produce the tubercles of rusted iron that broke free from iron pipe interiors and clogged valves, household faucets, and, perhaps most important, sprinkler systems for fire suppression. Because it was lighter and easier to install without heavy equipment than metal or ceramic pipe were, it could be used in locations with very porous soil, such as the nearly liquid clay near Tan Son Nhut airbase in Vietnam. Joints could be closed successfully with a minimum of leakage using only chains and a jack. Somewhat controversially, asbestos-cement pipe remains in use for sewers and water supply in many parts of the world, despite opposition from some environmental groups who fear contamination of drinking water.[68]

The last major new use of asbestos before health and environmental concerns began to reduce American consumption of the material was in pavement and curbing, one of the few end-uses in which its thermal and electrical nonconductivity were not significant factors. In 1961, the city of Atlanta's engineers began experimenting with asbestos asphalt and were impressed by the ability of the mix of 2.5 percent asbestos plus aggregate and other mineral filler to absorb 7.4 percent asphalt by weight, a high proportion, without flushing any asphalt to the surface. The resulting pavement was more flexible and less penetrable by water; it filled depressions in the old pavement underneath and could be applied more quickly than traditional bituminous paving materials. The city of St. Louis reported that it even resisted penetration by women's spike heels during the summer months. Moreover, in tests in New York State, its friction properties reduced the incidence of skidding on wet pavements by 30 percent.[69]

In 1963, the American Road Builders Association reported at its annual convention that "the use of asbestos fiber as an admixture for asphalt pavings increased by approximately 300% during 1962." The material was tested by the Naval Civil Engineering Laboratory in 1964, which did not think the mineral added any strength to the pavement, but it was nevertheless soon in use on the decks of barges and work boats. *American City* wrote enthusiastically of asbestos asphalt as an overlay on existing pavement after the Port of New York Authority resurfaced the upper deck of the George Washington Bridge with the material in 1968:

> Adding asbestos fibers to an asphaltic-concrete mixture boosts the amount of asphalt such a pavement can normally incorporate without bleeding, shoving and rutting under traffic. And increasing the amount of asphalt enhances several qualities including impermeability, an especially significant factor on bridges. . . . The resulting pavement is tougher, resists oxidation better, and will outlast normal pavements.

This last factor, durability in service, made asbestos asphalt cost-effective, although the initial cost of the material was higher than that of traditional mixes. As late as 1973, asbestos asphalt was still receiving rave reviews from municipal users.[70]

Asbestos was a successful pavement-mix additive in rubberized-tar concrete as well, passing tests by the U.S. Army Corps of Engineers' Waterways Experiment Station in 1959. The station concluded that asbestos increased weathering resistance, wear, deformation, heat, freezing, and jet-fuel penetration on aircraft runways. In the 1960s, the Delaware Department of Transportation specified "the incorporation of one to three percent asbestos fibers, by weight, in hot mixed, hot laid bituminous concrete curbing." The Delaware standard was revised in 1979 because of "the increasing health hazard associated with the use of asbestos, as well as shortages of supply."[71]

Through the 1950s, 1960s, and early 1970s, most of the consumption of asbestos in the United States was in transportation, especially friction materials, water and sewer pipe, and building construction, including asbestos-cement ductwork. Architects and engineers continued to develop new ways of using asbestos in building, particularly for low-cost housing in the United States and elsewhere; and in doing so they had the support of fire-safety professionals who, as we have seen, had plenty of historical experience with low-cost housing so combustible that it was astonishing that anyone would take the risk of living in it. The U.S. International Cooperation Administration's Technical Aids Branch and the United Nations Technical Assistance Board both endorsed asbestos-cement as a building material for developing countries.[72]

The fire-resistive qualities of asbestos in building materials had taken on a new importance in the cold war era. In the ten years after World War II, nearly every organization that published model building codes issued a new edition that reflected the lessons of the war, including not only incendiary bombing but the possibility of nuclear attack as well. States and provinces that had not previously adopted codes responded to President Truman's call for greater attention to fire safety at all levels of government and enacted or revised codes for construction outside municipal jurisdictions in their states. This meant, at the very least, that schools, hospitals, hotels, correctional facilities, and other assembly occupancies outside the fire limits of the nearest community were expected to comply with the state code and that all municipal codes were expected to be consistent with those of their states. As in previous waves of code adoption and revision, most of these new codes were based on the model codes published by organizations such as the NFPA, the Building Officials' Conference of America (now called Building Officials and Code Administrators), the National Board of Fire Underwriters, and the Southern Building Code Congress.

Between 1942 and 1958, according to the National Board of Fire Underwriters, a fire broke out in a community of more than 2,500 persons in the United States, on average, every thirty-seven seconds—about 812,000 fires per year. The largest number of fires occurred in 1952, when 983,733 fires were recorded. Although multiple-death building fires peaked in the United States during World War II, the number of deaths from fire actually increased in the postwar period, averaging 11,602 fatalities per year between 1951 and 1958.[73] The board worked with the NFPA to develop and publish model codes and handbooks of fire-safety practice based on consensus standards. The model codes of both organizations underwent frequent revision between 1945 and 1960, and the death toll from fire began to fall.

The NFPA's *Handbook of Fire Protection* came out in a tenth edition in 1948, an impressively thick octavo volume of 1,544 pages, heavily illustrated with charts, graphs, drawings, and photographs. This tome contains scores of references to asbestos as a thermal insulator under various kinds of heating devices; in furnace and boiler mountings and insulation; as a heat barrier behind, around, and under stovepipes and smoke pipes; around gas-vent piping; in ductwork; around steam and hot-water pipes to prevent ignition of wood floors and partitions; behind wall heaters; for the walls, floors, and ceilings of dry rooms and caul boxes; firestopping and insulation of chimneys and their flues; in fire doors and partitions; for welding curtains; as a filler for plastic resins; in electrical panels; in flooring; in stuccos and plasters; in roof coverings; as water pipe; and in gaskets for heated water tanks.[74] The NFPA's *National Fire Codes* for 1951 listed mainly performance standards without specifying materials; but for some applications, such as the flexible connections used in the construction of ducts, fire dampers in fire partitions, the insulation of ductwork where it passed through combustible materials, chimney baseboard linings, and insulation under heating appliances, asbestos was still recommended by name. Asbestos-cement and built-up shingles were, of course, among the approved roofing types listed in the model roofing ordinances, as were asbestos-insulated walls and partitions among the approved fire-resistive types.[75] Detailed recommendations for use of asbestos in fire-resistive building assemblies and industrial plants were also included in the 1954 (eleventh edition) of the handbook.[76] The *National Electrical Code*, also published by the NFPA, described seven types of asbestos-containing wire insulations, and provided detailed guidelines for appropriate uses.[77]

The 1955 edition of the National Board of Fire Underwriters' model *National Building Code* and 1960 *Fire Prevention Code* had the by-then obligatory description of the approved three-pound-per-yard asbestos safety curtain for theaters and asbestos board in motion-picture projection booths, plus asbestos concrete for fireplace hearth extensions; asbestos millboard under heating appliances, including both home and commercial cookstoves; as counter protection under small gas-burning appliances; and under any of sixty-nine listed types of industrial furnaces, heating appliances, and boilers. Steam and hot-water pipes enclosed in wooden boxes or covers "shall be lined with sheet metal of not less than 28 gauge or asbestos millboard not less than ¼ of an inch thick." Similar rules regarding asbestos and sheet metal applied to incinerators, except that both, or "protection equivalent thereto," were required. Exhaust ducts from paint and varnish spray booths were also to be insulated in this manner. Sheet asbestos was one of two approved materials for use as guards around welding and metal-cutting operations.

In residential and other warm-air and air conditioning systems, the board's code required that "[v]ibration isolation connectors in duct systems shall be made of woven asbestos or approved flameproofed fabric or shall consist of sleeve joints with packing of rope asbestos or other approved noncombustible material." For firestopping of supply ducts where they passed through walls, floors, or partitions, builders had a choice of asbestos, mineral wool, "or other noncombustible insulating material"; but vertical supply ducts had to be covered with "approved air cell asbestos not less than 1/4 of an inch thick" or some material that the builder or property owner could prove was equivalent in performance. Any firestopping removed for the installation of ductwork had to be replaced with approved firestopping as just described. Asbestos-cement shingles, built-up asphalt asbestos, and corrugated asbestos-cement roofing were all on the list of approved roofing materials. An appendix titled "Fire Resistance Ratings" cited ASTM E119, the predecessor of ASTM E84, and reminded readers, in the section on the limited fire resistance of plaster, that "[u]nless asbestos or other fiber is added to Portland cement plaster, its fire resistance is further limited by cracking and spalling." Among the ratings in hours of protection for "Beam, Girder and Truss Protections," "Floor and Ceiling Constructions," and "Walls and Partitions" were the specifications for approved assemblies for each, if asbestos was used as fire insulation. The 1968 building code of New York City appears to have been based primarily on this model.[78]

In its guidelines for evaluating the fire potential of municipalities, which, of course, affected the rates paid by fire-insurance customers in these communities, the National Board of Fire Underwriters gave very low marks to any jurisdiction foolish enough not to prohibit wooden shingles in its fire district. The board also published guidelines for its industrial customers, including those in the metalworking industry, who contended with metal dust and powder fires that could not be extinguished by water or foam and had to be snuffed out with large quantities of "dry sand, rock dust, ashes, talc, asbestos, and other similar inert powders." Asbestos was, as one might expect, one of the approved roofing materials for these customers.[79]

The Building Officials' Conference of America also mentioned asbestos in its model building code of 1951. Their code was in fact reprinted and endorsed by Connecticut's Public Works Department and distributed through the agency's Housing Division. The approved applications were predictably similar to those of the NFPA and the National Board of Fire Underwriters: asbestos shingles and cement boards were among the approved siding materials, furnaces should be jacketed "with 1½ [inch] asbestos cement," asbestos-cement board was an approved wall partition type, the material was one of the approved types in ceiling and floor assemblies, and it was required in the lining of chimney baseboards, where heated masonry could ignite woodwork.[80] The International Conference of Building Officials, which published a multivolume compendium of standards as its model *Uniform Building Code*, used asbestos as the zero calibration in its "Test Method for Fire Hazard Classification of Building Materials," number 42-1-67, and a version of E119-58 that the Building Officials' Conference of America called 43-1-67, the Steiner tunnel test based on asbestos for testing building construction and materials. Safes and vaults were also tested against the performance of asbestos in the International Conference of Building Officials' standards 48-1-67 and 48-2-67. The first volume of its code listed asbestos proscenium curtains as standard 6-1-67 and asbestos-cement shingles as 32-9-67. The other volumes in the set were essentially detailed restatements of the standards set forth in NPFA's and the National Board of Fire Underwriters' publications, including references to asbestos as an approved roofing, siding, and insulating material. Very little of this material was changed in or deleted from the 1988 edition.[81]

State codes of the 1960s were often based on performance tests for materials, but for some fire-safety applications asbestos was explicitly recommended. Nearly all states required asbestos and/or steel fire curtains in theaters, and most were quite explicit about the method of installation and the weight and type of fabric. In Massachusetts, asbestos was one of two materials acceptable for firestopping of vertical openings such as flues and ducts; New Jersey in 1965 required asbestos and/or twenty-eight-gauge metal under heating appliances, air-cell asbestos "or equivalent" around ductwork if not using asbestos-cement ducts, asbestos jacketing of furnaces, and asbestos and/or metal under low-heat furnaces and ovens. Radiators and heating coils in public garages and airplane hangars were to be protected with "asbestos or other insulation." For certain types of heating systems, there was no equivalent for "one-quarter (¼) inch asbestos millboard covered with No. 24 U.S. gage steel sheets under the appliances, projecting not less than eighteen (18) inches from the sides of the appliance where fired and where hot products of combustion are removed."[82]

The specification of asbestos with or without metal as thermal insulation under or around heating appliances, especially those burning solid fuel, was a persistent feature of building codes until the end of the twentieth century. The near-universality of this provision was remarked on in 1979 by engineer William H. Correale and by Richard Peacock of the National Bureau of Standards.[83] The latter was interested in the problem because the fire safety of wood-burning appliances had become, in the 1970s, a public policy issue through the association of these appliances with conservation of fossil fuels. A 1982 report by the National Bureaus of Standards for the U.S. Department of Energy and the Consumer Product Safety Commission remarked that "all the methods of wall protection specified in the model building codes use two materials: asbestos millboard in various thickness and sheet metal. With the current health concerns with the use of asbestos, few alternatives are left for wall and floor protection."[84]

But no handbook or code provides as panoramic a view of asbestos in the American industrial built environment as the Factory Mutual Engineering Division's *Handbook of Industrial Loss Prevention*, published in 1959 and 1967 in McGraw-Hill's Insurance Series. Factory Mutual, a mutual fire- and hazard-insurance company serving a primarily industrial membership, did not accept companies as members that were found at survey to be poor fire risks. To keep insurance premiums low for its membership, it maintained a laboratory and did its own testing of processes, structures, and materials to determine best practice from the safety perspective. In the handbook, directions for implementing these practices are usually expressed in the imperative mood; in the chapter on air conditioning, for example, we are told: "Use flexible woven asbestos or material of comparable fire resistance at flexible duct connections." Factory Mutual used many of the same test techniques as UL, ASTM, and NFPA, including those that employed asbestos-cement as the zero point of combustibility and flame spread.[85]

The first chapter of the 1959 edition concerns building construction, and it essentially reiterates the advantages of various types of fire-resistive construction and assemblies, and the rating in hours of each, that appear in other model codes. Chapter 2, "Preventing Horizontal and Vertical Spread of Fire," includes discussions of asbestos insulation in fire doors and asbestos paper and felt as components of the system of elevator and stair enclosures that permit flame- and smoke-free evacuation during a fire. In chapter 7, "Automatic Sprinkler Installation," we learn about the significance of iron or asbestos roofing in the choice of sprinkler type and placement and the need to insulate, with asbestos or an equivalent, the heaters required in cold climates in dry-pipe sprinkler valve enclosures. The perils of obstructions in sprinklers, including scale and the oxide tubercles from iron pipes are set forth, as are the advantages of using asbestos-cement pipe, which does not tuberculate, for the underground component of the sprinkler water-supply system. This virtue, plus asbestos-cement's light weight and ease of handling, is mentioned again in chapter 13, "Public Water Systems." Asbestos-cement pipe is one of the approved options for supplying fire pumps in chapter 15. "Water Tanks for Fire Protection," the subject of chapter 14, may have asbestos roof coverings fastened with "corrosion-resisting nails."

In the electrical section, in the chapter on "Generators and Switchgear," Factory Mutual warns:

> Avoid the use of slate panels; they are not in any case suitable for voltages over 750. They may have conducting metallic veins, and they have poor resistance to heat and shock. Ebony asbestos panels [asbestos in phenolic resin plastic] have high dielectric strength, insulation resistance, and ability to withstand shock, vibration, and rapid temperature changes.

The incombustible character of this material is described in greater detail later in the volume, in chapter 57. In chapter 26, we learn that asbestos-containing askarel transformers are more expensive than the oil type but safer and thus more cost-effective.[86] If our generator becomes wet, chapter 28 tells us we may dry it in an enclosure "of noncombustible material, such as corrugated iron, corrugated asbestos, or gypsum board." In chapter 29, "Special Electrical Installations," we are warned to use asbestos millboard or an equivalent between fluorescent fixtures and wood surfaces if the clearance is less than 3″, and are again reminded that approved transformers are of either the dry or the askarel type. We are offered no choice of materials for flameproofing grouped cables; these must be wrapped in the type of asbestos tape called "listing," applied "spirally with a 50 per cent lap," which is afterwards saturated with a sodium silicate solution with a specific gravity of 1.15.

This unstinting attention to excruciating detail continues in chapter 34, in which we are instructed on the means to avoid "Boiler Furnaces—Fuel Explosions," beginning with a "Safe Lighting-off Procedure" using "a reliable lighting torch," which "consists of a dozen turns of ½-in. asbestos rope, secured by a hairpin bend on a 5-ft. length of ⅜-in. diameter steel rod, kept immersed in a pot of light oil at the boiler front." If our furnace is coal-fired, it should have ⅛ inch of asbestos millboard between the inner and the outer cover plate of the register. In the next chapter, we learn that the gaskets for our "Heat Transfer by Dowtherm" "should be of spiral-wound steel and asbestos, solid Armco iron, Armco iron-clad asbestos, or nickel-clad asbestos." If we bury our liquefied petroleum gas tanks as described in chapter 38, we should protect them against corrosion with "one coat of primer, two coats of bituminous enamel, a layer of 15-lb bituminous-saturated asbestos felt, and a final layer of kraft paper." A few pages later in this chapter we are instructed: "Use gaskets of dead soft aluminum O-ring or spiral-wound stainless steel asbestos-filled types for tank-manhole, relief-valve, and pipe connection flanges" as well as in the distribution system. Chapter 39 tells us that the same layers of primer, enamel, asbestos, and paper used for liquid propane gas tanks will protect the steel pipes used in gas piping systems; we encounter this prescription again in the discussion of "Flammable-Liquid Pumping and Piping Systems" in chapter 44. A new (in 1959) type of gasholder is approved by Factory Mutual, which employs "no water, tar, or grease": "The dry-seal membrane has a core of woven asbestos glass covered on both sides with synthetic rubber." In the next chapter, we discover that we may house our acetylene generators in buildings sheathed in corrugated iron or asbestos-cement board.

Flammable compressors, chapter 40 asserts, should be in structures with venting windows of light corrugated asbestos or metal, as should the indoor storage of flammable liquids in drums mentioned in chapter 42 and that of materials subject to dust explosion, such as grain, in chapter 60. In chapter 40, we learn the approved equipment choices for electrolytic chlorine processes, including the Hooker-type cell in which "[t]he main body of the cell contains the steel cathode with hollow perforated fingers covered by an asbestos diaphragm, and with graphite anodes between the fingers." Chapter 44 advises us to use "spiral-wound or other metallic asbestos-filled gaskets of stainless steel, Monel, or copper" in our flammable-liquid piping systems and cautions against "compressed asbestos-fiber-sheet gaskets, since the organic binder used in the manufacture reduces their resistance to high temperature." The spray booths described in chapter 46 should have their ductwork sealed with "asbestos or mineral wool to prevent passage of flame or smoke." Carbon disulphide, we learn in chapter 52, should be stored under "sun roofs of light, noncombustible materials, such as corrugated iron, aluminum, or asbestos, on steel framing without side walls." Our distillery, should we be fortunate enough to have one, should be housed in a light structure as well, sheathed in "sheet metal or corrugated asbestos on steel framing, for maximum explosion-venting possibilities." We are given the same instructions regarding solvent processes in chapter 55.

Chapter 57 covers, in part, already-familiar warnings regarding the safe handling and storage of cellulose nitrate film. If we should be called upon to store rubber tires, racks are recommended in chapter 66, with barriers between the shelves which "may consist of tight ¾-in. plywood, ⅞-in. matched boarding, or firmly supported ⅝-in. flat cement-asbestos board." Single metal sheets are not approved. Our cooling tower can be made entirely incombustible with the guidelines in chapter 68, which include "a shell of galvanized copper-bearing sheet steel or asbestos wallboard." In the next chapter, we learn the best method for attaching our corrugated asbestos roofing, as well as the fact that "[s]mooth-surface roofing should have upper felts of asbestos"; no other material is approved in this application. No substitutes are acceptable in the case of the asbestos rope used in the "joint where sections of lining supported by shell of radial-brick chimney overlap." Finally, in chapter 74, we may minimize our risk from earthquakes by building "[o]ne-story steel-frame buildings having light wall and roof coverings, such as corrugated asbestos, corrugated iron, or asphalt-coated metal." Clearly, Factory Mutual's engineers and underwriters placed a great deal of faith in the virtues of asbestos.[87]

Because fire safety had become after 1945 a national security issue, codes, standards, and guidelines for best practice were rigorously tested in both the United States and other nations. Assemblies such as roof systems were tested as well as individual materials, not only by the traditional testing laboratories of UL, ASTM, and Factory Mutual but also by university laboratories such as those at the Massachusetts Institute of Technology, which had been a leader in fire research for most of the twentieth century. The National Bureau of Standards, which had, as we have seen, played a role in asbestos's emergence as a fire-resistive material early in the twentieth century and that had assisted in the development of federal building codes in the late 1940s, began a program of building technology testing in the late 1960s at the request of the National

Research Council. The bureau tested the flame-spread properties of interior finishes, including those containing asbestos; experimented with the role of interior finishes in fire buildup in rooms; and reinvestigated the efficacy of the Steiner tunnel test for determining flame spread of materials.[88] Fire resistance of buildings, especially the systems and assemblies used in high-rise structures, were also widely discussed in the international engineering and architectural communities, with increased sharing of information and building-fire experience from Europe, Asia, and the United States into the 1970s and 1980s. Asbestos was a standard material in these assemblies and systems.[89]

In Britain, which did not have the kind of sophisticated private testing facilities that ASTM and UL provided in the United States, the Department of Scientific and Industrial Research undertook both sponsored testing of materials and assemblies submitted by industry and its own tests of fire-resistive building materials. The agency published an authoritative list of such materials in 1959 and in subsequent years performed a series of tests on structural elements, of which twelve of those published in the 1960 report contained asbestos. When assemblies flunked the tests, as some inevitably did, this was duly noted. The Fire Research Station, under the jurisdiction of the U.K. Department of the Environment in the late 1960s and 1970s, went on to test the fire hazards of ceiling linings (including those that contained asbestos), the effects on fire spread of roof construction and contents, and the performance of asbestos fire blankets.[90]

In both Britain and the United States, government agencies were concerned about the national defense implications of combustible materials in the built environment. Britain, which had so little native timber that few buildings in the twentieth century had been built of it, had learned during the Blitz how seemingly innocuous elements in a structure, such as wainscoting and furniture, could be deadly to inhabitants in a serious fire. All combatant nations in World War II had learned, by either observation or experience, that even ordinary air-raid shelters under buildings are inadequate protection during a firestorm or a moving fire front, as occupants will asphyxiate when the fire consumes all the oxygen in its passage over them. When considering the possibility of nuclear war, the implications were grave: a one-megaton nuclear explosion would cause fire damage to every structure within a 660-square-mile radius, and it would be impossible to fight the resulting fires because of the danger from radioactive fallout within a half-hour of the burst. Fire maps of Hiroshima drawn after the attack in August 1945 clearly showed these effects (see fig. 5.4). It is also unlikely that streets would be passable for fire equipment after the blast. As Diane Diacon of the Building and Social Housing Foundation wrote in 1984, "spontaneous ignition of paper and fabrics will occur up to eight miles from a 1 MT air-burst explosion" during the heat wave, or second phase, of a nuclear explosion, causing not only ignition of combustible materials but third-degree burns to exposed humans.[91]

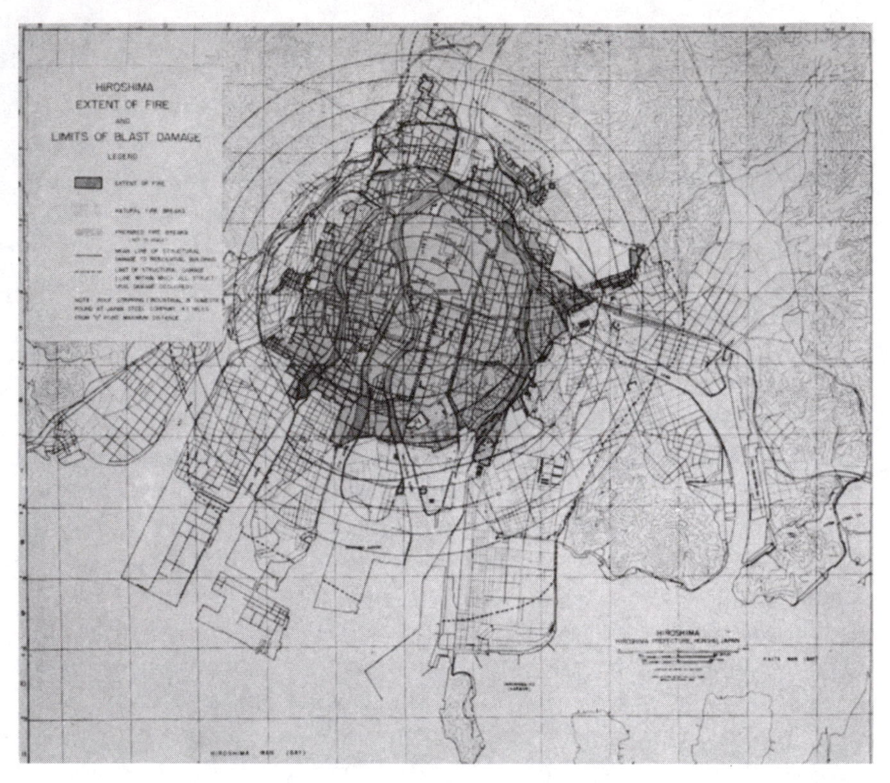

FIGURE 5.4 National Fire Protection Association map of the fire area of Hiroshima in 1945, made for the U.S. Strategic Bombing Survey. The radiating circles depict gradations of structural damage; the dark area in the center was completely burned out. The irregular dotted lines show fire damage to buildings as distant as four miles from air zero. The civilian-defense concerns raised by nuclear weapons drove significant changes in fire-safety policy after the war.

Horatio Bond, ed., *Fire and the Air War* (Boston: NFPA, 1946), 197.

Diacon, like her predecessor analysts of the effects of nuclear explosions on the built and human environment, relied on the work of Samuel Glasstone, who studied both the Japanese atomic bomb effects and those of earlier and later test detonations between 1945 and 1962.[92] Planners working from this and the Japanese volumes of the *Strategic Bombing Survey* concluded that American cities were unacceptably dense—"built-up" had been the term used by analysts of incendiary bombing—and too combustible to withstand a nuclear attack. They advised reducing combustible materials in construction, especially roofs, and spreading out the population over larger areas, preferably with wide firebreaks between neighborhoods. The effects of this theory can still be seen in the urban renewal of communities such as the downtown area of Elmira, New York, and the East Liberty business district in Pittsburgh: they are now much safer from fire and possibly also from thermonuclear effects but were economically devastated by the physical and social fragmentation of the neighborhood. Suburbanization was, of course, also encouraged by this policy of dispersal; and large numbers of houses of fire-resistive construction that included asbestos in walls, ceilings, floors, roofs, and other applications were erected outside urban centers. Nuclear planning was not the only force behind this trend, of course, but it justified as a national defense measure the flow of federal funding into urban renewal and suburbanization projects.[93] Manufacturing plants were encouraged as well to use materials, such as the corrugated asbestos-cement roofing and siding that had proved their worth in Hiroshima and Nagasaki, in new construction and renovation of older structures. In 1962, the Defense Department's Office of Civil Defense published a series of plans called "Family Shelter Designs," which included PSD F-61-2: "Basement Corrugated Asbestos-Cement Lean-To Shelter," which could be easily and inexpensively built to house a family of three in the event of nuclear attack (see fig. 5.5).[94]

FAMILY SHELTER SERIES **PSD F–61–2**

Basement Corrugated Asbestos-Cement Lean-To Shelter

FIGURE 5.5 U.S. Defense Department drawing published with plans for a "Basement Corrugated Asbestos-Cement Lean-To Shelter" in 1962. The homeowner was instructed to build this shelter against a basement wall, cover it with sandbags, and stock it with necessities like water and food, leaving one end open but stocked with additional sandbags. In the event of a nuclear alert, the family was to enter the shelter and close it with the stockpiled sandbags.

U.S. Department of Defense, Office of Civil Defense, *Family Shelter Designs* (Washington, D.C.: U.S. Government Printing Office, 1962).

Asbestos had been the perfect material for the industrial era, manufactured by 1970 into nearly 3,000 different products, most of which protected life and property from heat, flame, and electricity. It had properties that made it invaluable in warfare on the sea and in both incendiary and nuclear warfare from the air. It was used in virtually every industry from hotel keeping to chemical manufacturing and had an installed base in building construction that ranged from shacks to skyscrapers in every community in the United States (see appendix A). In the mid-1960s, the focus of this picture began to change from the benefits of the mineral to its risks and costs, slowly at first, then gathering momentum through the seventies and eighties through both litigation and government regulation. The process by which asbestos fell from grace is the subject of my last chapter.

6

The Asbestos Tort Conflagration

After 1965, in part due to the success of the fire-safety system that included asbestos, short-term fire safety moved into the background of discussions of the material, and its long-term health hazards were generally treated as if they were the only safety considerations associated with asbestos. The marker event for this change was the New York Academy of Sciences conference on asbestos in October 1964, the results of which were published the following year. Organized and led by Irving J. Selikoff of Mount Sinai Hospital in New York City, a charming, courageous, and compassionate medical professional with more charisma than credentials, a small group of physicians and attorneys began to work actively to alert workers, government agencies, and the general public to asbestos's character as a carcinogen. Selikoff had investigated a group of union asbestos workers in the early 1960s and obtained results that suggested increased risk of cancer for this group. This is the historical moment characterized by Paul Brodeur as the point "when Dr. Selikoff descended from Mount Sinai with his tablets of epidemiological truth."[1] A larger study by Selikoff and his colleagues was funded in 1968 by Johns-Manville and the International Association of Heat and Frost Insulators and Asbestos Workers, who must have been equally alarmed by the potential threats to their livelihoods posed by the papers presented in New York City in 1964.[2]

The question of who knew what when about the risks of asbestos has been set forth in such detail elsewhere that I have touched on only the aspects of this debate that are germane to my argument.[3] Although, as predicted, asbestosis rates had declined with workplace dust abatement through the twentieth century, mesothelioma and lung cancer cases began to appear in the 1930s and to increase thereafter, gradually but noticeably.[4] As we have seen, articles appeared in medical journals and industrial hygiene publications during this period. When the 5 million particles per cubic foot standard was proposed in

1935 and established in 1950, many medical professionals assumed the rate of malignant neoplasms would fall along with that of asbestosis.[5] The National Academy of Sciences asserted in 1971, "It was after 1960 that serious consideration was first given to asbestos as an etiologic factor in mesothelial malignancies."[6] By late 1964, it was clear to the occupational medicine community that the two diseases were not well correlated with each other and that it was possible to have mesothelioma without asbestosis and vice versa.

The long latency period of mesothelioma and lung cancer, the synergistic effect of smoking, and the persistent inability of health professionals to determine a maximum safe exposure to asbestos, even after 1965, confused the issue of compensation for injury. Many of the sick had worked for several employers and in more than one state, making workers' compensation claims difficult to document. In some states, workers could not sue their employers or former employers; others had statutes of limitations shorter than the latency period for asbestos-related cancer. Lacking the kinds of national health insurance and social support for occupational illness that European and other industrialized nations could rely on when asbestos-related diseases began to appear in their populations, sick workers and former workers could only fall back on America's traditional panacea for social problems: litigation.

The government agencies that had endorsed asbestos, and either implicitly or explicitly specified it in procurement, categorically refused to accept responsibility for the consequences of its use; indeed, the Environmental Protection Agency even attempted to ban asbestos altogether in 1989 and was only prevented from doing so by the finding of the Fifth Circuit Court of Appeals, among other objections, that it was not in the public interest to ban a material for which there were then no acceptable substitutes in applications such as brakes.[7] With the government refusing to accept liability, the only deep pockets that remained were those of the dozen or so companies that had formed the core of the American asbestos industry, and their insurers. One by one, the manufacturers declared bankruptcy as claims against them began to approach or exceed annual revenue at a time when markets were falling precipitously due to the perceived risks of using the product. Plaintiff attorneys began to look farther afield for potential defendants with enough money to make the effort worthwhile.

As the pool of possible defendants was expanded in the 1990s, so many firms began declaring bankruptcy that a monthly newsletter that tracks these developments, *Mealey's Asbestos Bankruptcy Reporter*, began publishing in 2001, put out by the same firm that has been publishing the biweekly *Asbestos Litigation Reporter* since 1993.[8] About 60,000 jobs associated with the bankrupt companies have been lost in the past decade as beleaguered firms have reduced themselves to shells with only enough staff to pay out from dwindling resources the claims of creditors, legal settlements and judgments, attorney's fees, and, in many cases, the pensions and health insurance plans of their retirees. The claims of the former all too often threaten to reduce the share of the latter to nothing. Only about thirty-seven cents of every dollar spent on asbestos settlements and judgments are actually paid to plaintiffs. At this writing (early 2004), more than 600,000 asbestos claims have been filed in the United States, of which the Rand Institute estimates that about 90 percent represent claims by persons whose health and fitness to work are entirely unimpaired but who must file before the closure of their states' statutes of limitations if they are to receive compensation for possible future illness.[9] Sporadically, Congress has attempted to devise some remedy for the worsening asbestos tort crisis, but so far these efforts have all gone down to defeat.

Various projections have been made in the United States and elsewhere as to the probable number of deaths to be expected from asbestos-related disease. Most recently (April 2004), the Environmental Working Group Action Fund has asserted that 10,000 persons per year are dying from asbestos-related disease, a figure reportedly based on a study of U.S. government data, although the CDC's figures are closer to 3,000 annually. In 1988, D. E. Lilienfeld, an epidemiologist at the University of Minnesota, published an article with three colleagues estimating that, by 2009, about 131,200 Americans would have perished of asbestos-related cancer, not including any of the projected asbestosis mortality. As of mid-2004, more than 80,000, or about two-thirds, of these deaths have yet to occur, suggesting that at least some of these projections err on the side of pessimism.[10]

The restructuring of the definition of strict liability in tort law set forth by the American Law Institute in section 402A of the *Restatement of the Law Second: Torts* in 1965 and the lifting of the ban on aggressive pursuit of prospective litigants by attorneys contributed to this frenzy of asbestos litigation, which has created economic problems worldwide. Lloyd's of London very nearly went bankrupt in the early 1990s when it could not close its accounting years with U.S. asbestos suits still pending as liabilities.[11] Hundreds of thousands more cases—up to a million—are expected to be filed in the next ten years as plaintiff medical and legal counsel assert that even one asbestos fiber in the lungs is sufficient to cause mesothelioma.[12] Given the quantity of asbestos used in the United States in the twentieth century—the "installed base"—there cannot be a single man, woman, or child living here who does not have at least one asbestos fiber in his or her lungs and thus no obvious limit to the potential number of plaintiffs. Physician Ronald E. Gots says of the situation that although "we customarily breathe in 1,000,000 or so asbestos fibers each year," no study has shown any significant risk of fatal or disabling asbestos disease to the general population. Any change on an X ray can, however, trigger a lawsuit, Gots observes, adding, "Claims for damages have pressed inexorably for a definition of asbestos-related disease at this subclinical extreme."[13] When a single building burns down, firefighters call it a building fire. When an entire neighborhood or city is incinerated, it is known as a conflagration. Without federally administered health insurance, tort reform, or some similar plan of the kind that the Rand Institute calls an "administrative solution," it is difficult to see how the conflagration of asbestos torts can be extinguished.

Why Asbestos?

One of the many important questions raised by the current political and legal tangle associated with asbestos is why this particular material has been selected as a toxicological, ideological, and economic battleground rather than any of the many other possible candidates for this dubious honor. As I have pointed out, we use many known toxins and carcinogens every day of our lives: our carpets and furnishings exude formaldehyde, most of our thermometers contain mercury, and lead is the material of choice for many industrial applications and indeed is still sometimes found wrapped around the necks of bottles of fine wine. Christian Warren has written a fine history of lead poisoning in the United States that makes it clear that lead is and has always been a hazard to human life. In 1978, according to the Environmental Protection Agency, 3 to 4 million children had "elevated blood lead levels."[14] Nevertheless, neither this nor other industry has been driven from the marketplace by a blizzard of lawsuits in the way that asbestos has; there is no Rand Institute report on lead toxic torts, for example, or pesticide litigation, although both have been significant lawsuit issues for years.[15] Gasoline and wooden shingles are notorious fire hazards; more than 40,000 of us die in automobiles every year. Even foodborne disease carries off more of us every year than do asbestosis and mesothelioma combined.[16] Why, in the presence of this cornucopia of potentially lethal hazards and thousands more just as bad or worse, have we chosen a single mineral to take the blame for all that some of us seem determined to believe is demonic and depraved about the alliance of technology and capitalism?

What accounts for asbestos tort exceptionalism is by no means obvious. The standard explanation is that the industry was guilty of "outrageous misconduct," to use Brodeur's classic hyperbole, in concealing, downplaying, or suppressing information about the hazards of its product, thereby committing "corporate malfeasance and inhumanity to man that is unparalleled in the annals of the private-enterprise system."[17] Leaving aside for the moment the hundreds, possibly thousands, of historical examples of much worse entrepreneurial behavior that are well known to historians, this argument fails to account for the uniquely deep economic black hole created by asbestos litigation, first because it is not unusual for companies to downplay product dangers, and second—I would argue mainly—because information about those dangers was readily available to anyone who really wanted to know from at least 1926 forward.

A third objection to this explanation is that there are significant examples of other industries that resisted for years or decades the association of their product with risk to workers and users and went on record in much the same way as asbestos executive Sumner Simpson did in his now-notorious papers, of which I shall have more to say later on. Lead paint is such an example; so is cotton dust, as I have already pointed out. Coal mines fought dust standards for years; in the case of coal, however, Congress accepted responsibility for the medical care of workers in an industry obviously vital to American security interests, and we have not seen the drowning of the coal industry under a tidal wave of liability as we have with asbestos. Gustave Shubert of the Rand Institute said of asbestos in 1985, when the number of cases was still only about 30,000, that "transaction costs as a whole are high, that is, higher than for any other type of tort litigation for which costs have been estimated."[18]

It has been argued that litigation has become a growth industry in asbestos because mesothelioma is so rarely associated with anything except exposure to asbestos; this is essentially the case that sociologist of science Sheila Jasanoff makes for the success of plaintiff medical expert witnesses in asbestos cases.[19] But other types of occupational illness, such as silicosis, are just as medically distinctive and are caused or aggravated by materials comparably as prevalent in the environment as asbestos. Yet no other toxic material litigation has had the massive destabilizing effects on the international economy that asbestos has. Another argument that can be entertained here is that asbestos, unlike lead and mercury and, for that matter, wooden shingles, is a relatively new product in the mass marketplace; but this fails to explain why even more recent products known to be toxic, such as pesticides, have not generated comparably devouring firestorms of product liability torts.

The only explanation that cannot be readily demolished with counter-examples is that a few key cases in the asbestos litigation history, particularly *Borel v. Fibreboard* in 1973, made it apparent that the legal profession had an unprecedented opportunity in the relatively helpless situation of American workers and former workers with respect to health care for catastrophic illness. The courts were ready to accept, at least half the time, a set of arguments that after 1984 could be prepared as if from a recipe, using ingredients listed in a readily available cookbook: Barry I. Castleman and Stephen L. Berger's *Asbestos: Medical and Legal Aspects*.[20] Both plaintiff and defendant attorneys belong to organizations that share information, including evidence, about ongoing litigation. The success of this just-add-local-circumstances-and-serve formula has been nothing short of remarkable, and it is hardly surprising that such a feast would attract large numbers of plaintiff attorneys who, like the rest of us, have to eat.[21] The Supreme Court ruled in 2000 that juries could award damages even for the fear of asbestos-related cancer, in the absence of any medical evidence of harm. As we shall see, the frequency with which plaintiff lawyers and their clients are rewarded with large sums of money, either in settlements or in jury awards, is even more surprising in light of the questionable validity of some of the arguments.

Nine Dead Mice: The "Conspiracy of Silence" Hypothesis

It is an article of faith in the anti-asbestos literature that the dangers of asbestos disease were deliberately kept from all but a privileged few, thus preventing workers from understanding the risks they took in the workplace.[22] This is in fact a significant component of the tort (the wrong) supposedly committed by the hundreds of businesses and ex-businesses now facing or having already faced asbestos disease claims so large as to carry into bankruptcy whole networks of international enterprise, most of them, at this point, from unimpaired plaintiffs. Yet these risks were apparently so well known as to have been the subject of more than seven hundred articles published before 1965. A conspiracy of silence that results in seven hundred publications surely cannot be considered a success, as a spokesperson from Johns-Manville pointed out in 1978.[23] One has only to consult the *Industrial Arts Index* volumes after 1926 to observe that the risks of handling asbestos are treated in the engineering literature in ways similar to those of lead, mercury, and other materials with known hazards. Articles are listed under the heading "Asbestos," most of them on engineering and geological topics; under "Asbestos Workers" and later "Asbestosis" are listings of recent articles on health research on the material, just as there is for other hazardous materials.[24] If indeed there was an attempted conspiracy of silence regarding the health risks of asbestos between 1930 and 1965, it seems to have been a resounding failure.

Anyone with an interest in the subject, which apparently did not often include the staff or elected representatives of the insulators and building trades unions, could have learned about these risks as workers in other occupations learned about lead, mercury, gasoline, and other known hazards. An hour in a public library at any time after 1926 would have revealed all of the "secrets" the asbestos products manufacturers were allegedly trying to conceal.[25] British historian Geoffrey Tweedale is in fact quite critical of the failure of British trade unions to take this obvious step and assume a leadership position on worker safety in the asbestos industry.[26] Their American counterparts do not seem to have been any more eager to rock the boat on which the livelihoods of their membership depended; there is no evidence that any union in the United States undertook even a routine literature search on the health hazards of the asbestos-containing products their members handled before Irving Selikoff began studying asbestos disease in the 1960s. The union magazine *Asbestos Worker* officially advocated the use of respirators in October 1944, but illustrations in later issues almost invariably showed workers handling asbestos without them. If union leaders were hesitant to bring to light whatever they may have known about the dangers of inhaling asbestos for fear of destroying the industry that put food on their members' tables, subsequent events have certainly shown that their concerns were more than justified.[27]

It is, of course, argued that manufacturers of asbestos-containing materials should have put warning labels on them so that users would recognize the risks they were taking when handling them and, one assumes, don dust masks despite the discomfort or perhaps change industries. This is asserted despite the well-known and amply documented fact that workers of any kind understandably dislike masks and respirators and will rarely wear them unless made to do so. Repeated warnings of danger do not seem to affect this picture very much; only threats of discipline and dismissal appear to result in consistent use of safety gear of all kinds. Equally well known is that no matter how warnings are expressed, a significant proportion of the target population will ignore or discount them, as in the case of automobile seat belts.[28]

Some argue that all the dust should have been removed from the work environment so that workers would not have to wear respirators. It is unclear what technology might have been available that would have been capable of accomplishing this, particularly in the case of products such as lagging and 85 percent magnesia insulation, which were often installed on a contract basis in workplaces controlled by their owners, not by the contractors. If there was a technology that would have entirely freed an asbestos-handling facility from dust in the period between 1930 and 1980, it is odd that the engineering literature does not mention it in the many articles on dust abatement. The National Academy of Sciences' 1971 report on asbestos air pollution asserted that for removing dust from some types of confined spaces "no satisfactory equipment is available."[29]

As for warnings on products, these were neither customary nor required before the era of the Occupational Safety and Health Agency (OSHA), which began in 1970. It was so unusual to place warning labels on products before the OSHA requirement in 1970 that even hundred-pound paper bags of ammonium nitrate, the explosive that leveled Texas City in 1947, killing 516 and injuring more than 3,000, did not then have to carry notices that the material could be explosive in a fire.[30] Meat, fish, and poultry have always harbored microorganisms that can make us sick or kill us, if we do not kill them first by heating the product; but only since the 1980s has meat, fish, and poultry packaging identified the raw contents as a hazard. We do not on this account see charges of conspiracy against, for example, the Beef Council of America or the U.S. Poultry and Egg Association on the grounds that its members concealed or downplayed the dangers of fresh product, although it is surely obvious that for reasons of self-preservation they would hardly have wanted them shouted from the housetops. It is a rare industry that would not prefer that its products appear safer than it is really possible for most products to be. As Groff Conklin wrote in 1949, "The reluctance of industrial managements to publicize the existence of a carcinogenic hazard is understandable, for it could easily become difficult for plants so labeled to recruit workers, or even to sell their commodities."[31] Very easily, as we have seen.

Indeed, one might argue that the "conspiracy" to conceal the dangers of wooden shingles has been far more successful than the one alleged to have existed in the asbestos industry: homeowners in fireprone California and other forested areas are still happily nailing these known fire hazards to their roof sheathings, despite an average of more than 9,000 wooden roofing fires per year in the United States between 1980 and 1998—nearly twenty-five per day.[32] It is difficult to understand why the self-contradicting and ultimately absurd argument for a conspiracy in the case of asbestos has been so frequently accepted in courtrooms.

The conspiracy argument seems to rest on a small group of documents relevant to two sponsored research projects between 1943 and 1950, both involving the kind of tiny data sets that characterized the medical literature of asbestos disease before 1965. The first concerns Leroy Gardner's serendipitous finding in 1943 that of eleven mice dusted in an animal experiment on asbestosis, nine developed cancers. Gardner quite appropriately expressed the view that this result suggested that additional research was needed on the carcinogenicity of inhaled asbestos. His corporate sponsors declined to fund this research and were apparently reluctant to allow him to publish his results.

Meanwhile, others, such as E. J. King and K. M. Lynch, were in fact carrying out experimental research on cancer in rodents exposed to asbestos dust and publishing their results. King et al.'s research on rabbits was published in 1946. K. M. Lynch et al. used mice; W. C. Hueper used rats and rabbits; and Arthur J. Vorwald et al. used rats, rabbits, and guinea pigs as experimental subjects in the 1950s; in the following decade A. G. Bobkov, J. C. Wagner, and G. E. Westlake, among others, published results of experiments with the effects of asbestos on rats and J. M. Jagatic, F. J. Roe, and R. L. Carter on mice.[33] It is difficult to see how much further contribution Gardner's proposed research project would have made to the state of medical knowledge, especially as it is unlikely he could have completed it before his death in 1946. Furthermore, research sponsors both private and public routinely turn down funding proposals, and private sponsors equally routinely regard the results of those they fund as proprietary, regardless of the subject matter.[34] The mere fact of their doing so is no more unusual or inherently sinister than the privileging of communications between attorneys and their clients and doctors and their patients.

Animal studies are in any case notoriously inconclusive as a means of determining a substance's toxicity to humans. Vorwald and John W. Karr, writing in 1938, and Vorwald with T. M. Durkan and P. C. Pratt in 1951, observed that different species of animals had quite different responses to asbestos exposures.[35] Even within a single species results could be confusing. In animal inhalation experiments at Saranac Laboratory, dogs did not show "peribronchiolor fibrosis of the lung similar to human asbestosis after being exposed by inhalation of intratracheal injection to long chrysotile asbestos fibers," according to Vorwald et al.; but Philip Ellman published a report of a dog that lived near an asbestos plant in South Africa in the 1930s that had impaired lung function.[36] Pigs are sufficiently similar to humans to make them good specimens for dissection by medical students, but they can ingest without harm substances that would kill one of us. Chocolate can be fatal to dogs and cats; any suburban or rural resident of the deer-infested northeastern United States knows that these ungulates can eat all sorts of ornamental and useful plants that are toxic to humans. The results of animal studies must, therefore, be confirmed by studies on our own species, as some of the asbestos-disease researchers have pointed out.[37]

The second supposedly suppressed research finding on asbestos-related cancer was that of Gerrit Schepers, who wrote a thesis at New York University in 1949–50 in which he reported eleven cases of pulmonary cancer among Canadian asbestos mill and mine workers. He later (more than thirty years later) testified that he had been pressured by Dr. Anthony J. Lanza of the Institute of Industrial Medicine at the university not to publish the results of his research and that his report to the South African government on his findings was "suppressed"—that is, not published. Again, this was one study purportedly lost from a body of literature consisting of hundreds of studies; Schepers does not tell us why he yielded to Lanza's pressure and why he did not simply do as any scientist might and submit an article on his findings to a refereed journal. It is difficult to see how either Lanza or the South African government could have prevented such a publication. This case and that of the nine dead mice constitute nearly the entirety of the support for the "conspiracy of silence" argument. The only other evidence is a set of letters written in the 1930s in which an asbestos company executive, Sumner Simpson, and an industry magazine editor opine that it would be best not to publicize the risks of asbestosis in the magazine *Asbestos*, much as the wooden shingle industry executives had expressed the desire a little earlier in the twentieth century that the dangers of their product should not be prominently discussed by fire-protection engineers.[38]

Even if one is willing to postulate some sort of attempted conspiracy on the part of some firms in the asbestos industry—clearly an unsuccessful one, judging by the quantity of published data—the questions raised about who knew what about asbestos and when they knew it are ultimately irrelevant. Even if everyone concerned then had known, say, in 1955, exactly what they know now, I still do not believe we would have discontinued the use of asbestos before the rate of fire deaths was reduced to an acceptable level, nor would it have been reasonable to do so.[39]

By 1955, the medical literature had reported some ninety-nine cases of asbestosis and cancer of the lung. Thus, the international community of experts on health and safety knew of fewer than a hundred cases of asbestos-related disease worldwide over a fifty-year period, to weigh in the balance against the deaths of more than 10,000 persons a year by fire at that time in the United States alone.[40] As a technological trade-off, this what is known colloquially as a no-brainer. Some of the safety experts who contributed to the writing and enforcement of building codes must have been aware of the literature on asbestos-related disease—they were, after all, safety experts—but in the face of undeniable evidence that the system of safety developed between the 1870s and the mid-twentieth century was saving more lives every year, these professionals evidently saw no reason to remove from it a component that had proven its worth as a fire-resistive material in hundreds of safety-critical applications. Indeed, engineers expressed deep concern about fire safety when public pressure led the Environmental Protection Agency to consider banning asbestos in the late 1970s and 1980s.[41] Asbestos use in the United States had reached record levels in 1973; in 1975 world demand still exceeded supply. Obviously, many engineers and fire-safety professionals still believed in the product.[42]

The demographics of the two kinds of mortality would have made the choice that much more obvious: nearly 40 percent of the fire deaths were children, while nearly all of the asbestos-related disease cases were and are adults. Due to its long latency period, mesothelioma is in fact typically a disease of older adults. For parents sending their children in, say, 1959 to what were widely known as firetrap schools, the possibility of disease in a relatively small number of adults three decades later could not have been anything like as lively a concern as that produced by newspaper photographs of the rows of small coffins interred after the fire at Our Lady of the Angels (see fig. 6.1). Almost as many persons perished in that one fire as in all of the fatal cases of asbestos-related disease reported in publications by that time, and all but two of the OLA dead were children.[43] The Cocoanut Grove fire claimed five times as many victims in twelve minutes as asbestos processing had in half a century.[44] While the illnesses and deaths from asbestosis and mesothelioma were and are certainly personal and family tragedies for the individual patients, they have been few in number compared with the victims of fire. Indeed, some of the persons now diagnosed with mesothelioma may well have lived long enough to develop it because they did not die in a fire. Because of the demographically younger profile of fire victims, in actuarial terms, the number of years of life lost to them are correspondingly greater than those who died of mesothelioma in middle life or later. Fifty years ago, 3,000 to 4,000 children a year died in fires; now 2,500 to 3,000 adults die annually of asbestos-related diseases. It seems doubtful,

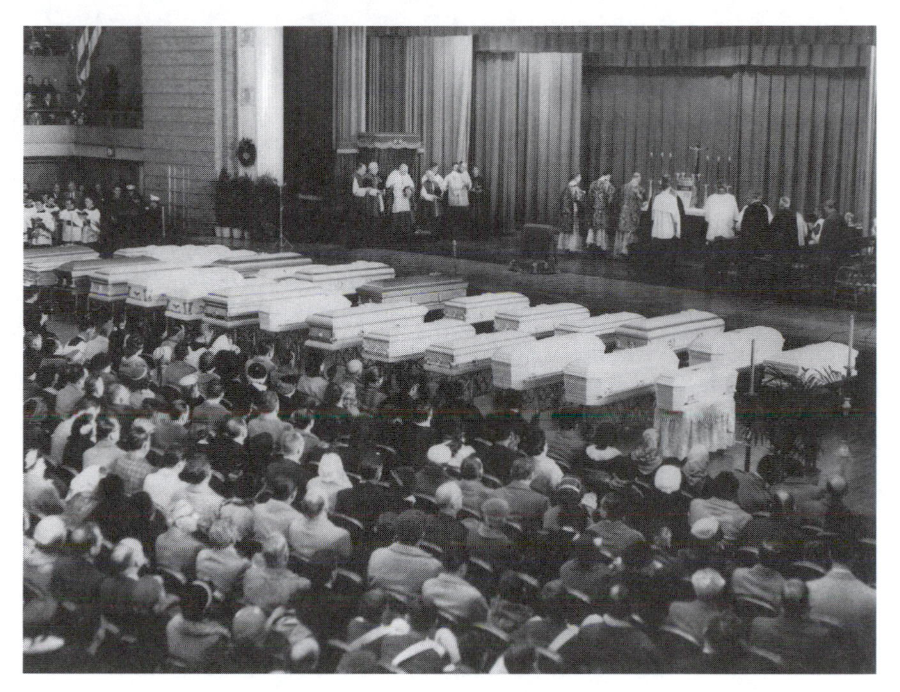

FIGURE 6.1 Funeral mass for twenty-seven of the ninety-five victims of the Our Lady of Angels fire in Chicago, December 5, 1958. The mass was celebrated in the Illinois National Guard Armory to accommodate the enormous crowd of parents, relatives, and other mourners.

Photo courtesy of the *Chicago Tribune*. Reprinted by permission.

even had all risks been known at the time, that the grieving families of the victims of the Collinwood school fire, or those of Our Lady of the Angels, Cocoanut Grove, the Rhythm Club, or the Winecoff Hotel, would have been much impressed by the human health implications of Gardner's nine dead mice.

Some writers also argue that the mineral was used because it was cheap, implying that businesses who used it in their work were cutting corners and acting improperly in so doing by endangering the health of their workers.[45] As we have seen, it is not the case that asbestos was cheap; most of the objections raised to its inclusion in construction codes, including those for shipping, related to the generally greater cost of asbestos compared with other materials. I have already discussed the significance of this in the case of roofing materials and ceiling tiles and provided comparison figures. It was endorsed by and included in construction codes by literally thousands of engineering experts concerned primarily with the safety of human life and with the conservation of energy, still a finite resource and a national security concern. Asbestos won the endorsement of these groups of professionals by superior performance, in tests performed all over the industrialized world, in comparison with most of its competitor materials in the applications for which it was endorsed in building and marine codes.

The question of testing is also systematically misrepresented in anti-asbestos literature, which routinely argue that asbestos was put into place without adequate product testing. This, as we have seen, is manifestly false. Asbestos was tested by safety organizations such as Underwriters' Laboratories, by the Department of Commerce's Bureau of Standards, by the U.S. Navy, by fire and marine engineers, and even by some school districts, most of which were almost exclusively interested in the immediate risk of fire to their constituencies and stakeholders.

The Medical and Legal Debate: Suffering Fools Expensively

Given that at least $250 billion is potentially at stake in asbestos litigation in the United States alone, plus $50 billion already spent on asbestos abatement and more than $1 trillion in depreciation of property containing asbestos, the intellectual rigor of legal discourse about asbestos seems remarkably lax. One would suppose that the large sums involved would motivate careful research and documentation of arguments, but one would, in fact, be wrong. On both the plaintiff and the defense sides of the debate, the level of discourse is not much more sophisticated than "You knew!" "No we didn't!" "Did so!" "Did not!" and so on. Efforts to introduce larger issues and examine the benefits as well as the risks of the technology in question have been largely ignored, particularly because most of these efforts have been condemned by the fallacy of origin: that is, the claim that they were made by industry executives.

Egregious and puzzling errors have been made on both sides. For example, only in early 2003 did it come to light that asbestos-disease crusader Irving Selikoff (1915–92) never in fact earned a medical degree, although he repeatedly insisted that he had done so. In the three decades or so that Selikoff was alive and testifying as an expert witness, defense attorneys apparently failed to notice this potentially explosive defect in his credibility. The issue of fire prevention has been raised from time to time by defense attorneys and witnesses, but the body of evidence—particularly the mortality statistics—have apparently never been systematically presented as a coherent published argument. This is a considerable oversight for a large group of presumably motivated persons who stand to lose billions of dollars. More recently, the integrity of some of the radiologists who have been interpreting plaintiffs' X rays has been called into question by a study at Johns Hopkins University.[46]

As to the origin of the problems in asbestos manufacturing, former Johns-Manville executive Bill Sells points out some of the ways in which the industry in which he was once employed shot itself catastrophically in the proverbial foot. In his critique of Johns-Manville published in the *Harvard Business Review* in 1994, he asserts:

> In my opinion, the blunder that cost thousands of lives and destroyed an industry was a management blunder, and the blunder was denial. Asbestosis—a nonmalignant lung disease brought on by breathing asbestos fibers—had been known since the early 1900s, and the first indications of a connection between asbestos and lung cancer appeared in the 1930s. But Manville managers at every level were unwilling to believe in the long-term consequences of these known hazards. They denied, or least failed to acknowledge, the depth and persistence of management accountability.
>
> Had the company responded to the dangers of asbestosis and lung cancer with extensive medical research, assiduous communication, insistent warnings, and a rigorous dust-reduction program, it could have saved lives and would *probably* have saved the stockholders, the industry, and for that matter, the product. (Asbestos still has applications for which no other material is equally suited, and correctly used, it could be virtually risk-free.) But Manville and the rest of the asbestos industry did almost nothing of significance—some medical studies but no follow-through, safety bulletins and dust-abatement policies but no enforcement, acknowledgment of hazards but no direct warnings to downstream customers—and their collective inaction was ruinous. [italics in the original][47]

The plaintiff side of the argument is not much more edifying: in chapter 2, I mentioned the propagation of error regarding knowledge of asbestos in antiquity from Castleman through later sources.[48]

This is only the most obvious and egregious example of carelessness about sources and arguments in the anti-asbestos legal, and sometimes even the medical, literature. As I noted in chapter 1, Stephen Berger's essay, "Alternatives to Asbestos Insulation," in Castleman and Berger's *Asbestos* makes the claim that "[p]atents contain the history of technological advances," as if a list of reported inventions were all one needed to create such a history. In seventy-seven pages (including the source notes), this essay omits all mention of the National Fire Protection Association, the Underwriters' Laboratories, the National Board of Fire Underwriters, the International Convention for the Safety of Life at Sea, the American Society for Testing Materials in general and ASTM E84 and its predecessor standards in particular, or the application of the standards of any these organizations to either municipal or maritime construction codes and standards, to say nothing of the existence and enforcement of international codes or those of other nations. Indeed, in reading the legal literature of asbestos generally, one would scarcely guess that any such organizations exist or that governmental units and the insurance industry exercise any regulatory power at all over the built environment.[49]

The argument from patents fails because only about 10 percent of patented inventions are ever produced and only about 5 percent ever return even the cost of patenting them. Berger is silent on how many of the materials that appear in his list "Alternatives to Asbestos Insulation" were ever actually available in the marketplace; and perhaps more important, he does not mention the necessity for these "alternatives" to meet the standards for low flame spread in building codes.

It has been argued that the federal government did not specify asbestos in military applications in World War II, despite copious documentation that it did so, both explicitly in conservation orders and specifications and implicitly through the use of performance standards such as ASTM E84 and its predecessor standards, and that these organizations propounded standards of performance that did not specify materials.[50] In some cases this is true; but in the relevant applications, the standards of performance were based on the performance of asbestos. In the case of flame spread, as we have seen, all other materials were judged by how well or poorly they performed compared with asbestos-cement.

No Free Lunch: Trading Off Risks and Benefits

Since our environments, natural and built, always have and always will contain irreducible risks to the body, the question was and still is whose bodies we will put at risk. Often, as in the case of asbestos, we do not know in advance whose bodies we are endangering, even, or perhaps especially, with safety technology, when we, for example, install airbags on the passenger side of automobiles or buy flame-retardant clothing fabrics.[51] We cannot be sure in advance what trade-offs we are making. This is the fundamental issue raised by the inclusion or removal of asbestos from the built environment, where it has formed part of our protection from fire for more than a hundred years. We cannot eliminate risk; but as our knowledge evolves with the use of a technology, we can at least attempt to decide which risks we prefer. As I have noted, we accept the deaths of tens of thousands of persons every year, not to mention the environmental costs, in exchange for the benefits of the automobile. We do not yet know how many additional fire deaths we will have to accept without asbestos in our built environment, and will not know for decades because so much asbestos will remain in place. Only about 2 percent of the built environment is being built or renovated at any one time. Moreover, we really do not know whether the alternatives are safer because we do not have the experience with great quantities of them in the built environment that we have had with asbestos.[52]

One of the most poignant—and ultimately unanswerable—historical questions associated with the trade-off between the risks and benefits of asbestos is that of its role in the World Trade Center. While the north tower was being built in September 1969, Irving Selikoff and two of his colleagues met with the contractor to express concern about the potential health effects of the sprayed-on asbestos fire insulation being used to protect the steel. After consulting with Selikoff, the Port of New York and New Jersey Authority agreed not to use asbestos above the fortieth floor of the north tower, and none was used on the south tower. Most of the asbestos that was used in the north tower was removed after the 1993 bombing of the Trade Center and replaced with a ¾-inch sprayed-on layer of cementitious mineral wool, which was in the process of being upgraded to 1½ inches in late 2001. The relatively low asbestos dust counts made in lower Manhattan after the disaster are consistent with this account. When the aircraft crashed into the buildings, most of the thermal insulation was blasted off the steel structural members on the impact floors, subjecting them to temperatures above 1,200 degrees Fahrenheit, at which point steel loses half its strength. The sprinkler systems did not operate because the water lines had been severed by the aircraft and the debris of the crash. Although the insulated spandrels and girders of both towers were rated for three hours of fire resistance, neither tower remained standing even for two hours. The Federal Emergency Management's 2002 report asserted:

> The fire safety of a building is provided by a system of interdependent fire protection features, including suppression systems, detection systems, notification devices, smoke management systems, and passive systems such as compartmentation and structural protection. The failure of any of these fire protection systems will impact the effectiveness of the other systems in the building.[53]

In the days and months after the attack, questions were raised about whether the asbestos insulation originally specified for the towers would have protected the structural steel longer and allowed more people to evacuate the buildings before the collapse. Most engineers agreed that nothing would have prevented eventual collapse; as Thomas Eager and Christopher Musso expressed it, "No designer . . . anticipated, nor should have anticipated, a 90,000 L Molotov cocktail on one of the building floors."[54] Nevertheless, some pointed out that even a small margin of extra time would have saved hundreds more who perished under the rubble. Whether asbestos would have provided this extra evacuation time cannot now be determined. One of the footnotes to the September 11 story is that fire-resistive accordion fire doors, described by a Polish visitor as "asbestos curtains," functioned as fire barriers during the attack on the Pentagon exactly as designed, closing off the long corridors that would otherwise have served as flues for the fire.[55]

The people of the United States made the decision to use asbestos collectively, as a democratic nation, with our eyes open but focused on immediate and not long-term dangers; we must now collectively take responsibility for the long-term consequences of these choices, as every other industrial democracy has done. We made these choices through our elected local and state governments after responsible safety professionals, who had access to the information regarding both fire and asbestos dust hazards, tested asbestos and recommended it in codes and standards. Thousands of organizations participated in these individual and local decisions: parent-teacher organizations, school boards, building committees, city councils, fire departments, women's clubs, and civic improvement organizations. Just as we support the health needs of our military veterans all their lives because they have faced dangers and taken risks to which we as a nation have exposed them, so should we accept our responsibility, as other nations do, to those who now pay the price for our technological decisions. No technology is a free lunch. Chauncey Starr wrote in 1969 that "implicit in every nonarbitrary national decision on the use of technology is a trade-off of societal benefits and societal costs."[56] Over the course of the twentieth century, Americans have accepted the tax burden of regulating, inspecting, and monitoring workplaces for safety, struggling at times to make adjustments for changes in technology. All the world's other industrial democracies have taken the next step in social responsibility for health care by offering universal coverage for their citizens.

We have all received the societal benefits of asbestos as fire-resistive material, and we are all responsible for the costs of it as a toxic substance. Moreover, some of the very persons now suffering from asbestosis and mesothelioma have received the benefits of asbestos in the form of protection from fire; Mary Douglas and Aaron B. Wildavsky point out that benefits as well as risks can be "hidden, irreversible and involuntary," making it difficult for any one individual to see how he or she has been affected by both.[57] We have all been exposed to asbestos and are therefore all, at least theoretically, at risk from it. In the United States, taxpayers and consumers are already paying the costs of asbestos litigation in everything we buy as well as in lost jobs. It would be simpler and more humane to ensure that all Americans receive the health care they need without requiring them to call a lawyer who can find somebody to blame for their condition.

The ambiguous and ambivalent quality of technology, and the necessity for making trade-offs in our decisions about using it, is the subject of Edmund Russell's book on DDT, *War and Nature*. Like asbestos, DDT was successfully used to save military lives in World War II. Russell points out: "Early in the Pacific War, malaria caused *eight to ten times* as many casualties as battle did."[58] A million children worldwide still die every year from malaria, making it difficult to argue for a worldwide ban on the pesticide. DDT was the subject of Rachel

Carson's *Silent Spring* in 1962, the publication of which was an important marker event for a change in public attitudes about the hidden and chronic dangers, to humans and their environment, of many technologies. The long-term effects of radiation, which had by that time been in the news for a decade and a half, contributed to these concerns.

The trade-off between fire risk and that of cancer is more difficult to assess than that between DDT and malaria, however, because malaria is an infectious disease with a known disease agent and a well-recognized vector. Cancer, as Robert Proctor makes clear in his 1995 *Cancer Wars*, is difficult to prevent in part because its causation is so complicated that it is often impossible to determine which of many possible exposures to carcinogens was responsible for the development of a particular cancer and how much of a role genetic susceptibility plays in an individual's response to any given dose of the carcinogen. A large number of "natural" materials, such as spinach and peanut butter, produce carcinogens in the body, but other components of these same foodstuffs help protect against cancer. Mesothelioma is usually, though not always, linked to asbestos exposure; but the number of cases of this disease is thought to be small compared with those of lung cancer in which asbestos might possibly be implicated. Cases of lung cancer, which have clearly been increasing since 1950, do not announce the identity or identities of their triggering carcinogen(s): the fatal insult to the original cell may have been delivered by tobacco smoke, X rays, radon, steel-mill dust, asbestos, or any of dozens of other possibilities. Moreover, cancers have long latency periods that separate the experience—and often the memory—of exposure from the appearance of the disease, further complicating efforts at prevention by analysis of causation.[59]

In response to slow-developing threats like that of cancer, some elements of the environmental movement have taken on, as Douglas and Wildavsky point out, many of the characteristics of eighteenth- and nineteenth-century pietism: "Infiltration from the evil world appears as Satanism, witchcraft, or their modern equivalent—hidden technological contamination that invades the body of nature and of man." In this model, it is always "they" (the external evil) rather than ourselves who have perpetrated the pollution.[60] The asbestos industry seems to have been an unusually convenient scapegoat of this kind.

But in fact, no one forced us or deceived us into using asbestos; and it is irresponsible, barbaric, and unacceptably costly to require sick persons to resort to litigation to recover their medical expenses. As the director of Massachusetts General Hospital said in 1942 of the medical expenses of the Cocoanut Grove fire victims, "It is a community misfortune, and its toll should be paid by the community at large."[61] For not much more than the $250 billion that the Rand Institute believes may be the ultimate price tag for asbestos claims, plus the $1 trillion in building depreciation we have already absorbed, we could have a national health insurance program that protects both workers suffering short-

and long-term effects of occupational hazards and the jobs created by companies now being driven into bankruptcy.[62] We are already paying these costs in everything we buy, except that now half of the proceeds of asbestos litigation go directly into transaction costs—that is, mainly to lawyers—and more than half of the costs of health care go into paperwork.

All democracies recognize that national security depends on the health of every citizen. To paraphrase Air Marshal Slessor, whom I quoted in chapter 4, our first responsibility is to keep ourselves alive and free, including the freedom from fear that some illness or injury, however caused, will deprive us and our heirs of everything we have. Making technological decisions is an important part of our collective freedom; accepting the consequences of them to national health is an equally important part of our collective responsibility.

APPENDIX A. SOME ASBESTOS END-USES IN THE UNITED STATES, 1850–1990

Acoustical tile and panels
Adhesive
Air-brake cylinder-packing expander rings
Aircell insulation
Aircraft hangars
Air-pump packing
Appliance insulation (as in electric toasters, hairdryers)
Aprons
Asphalt mastic flooring
Automotive undercoating

Bags
Baking paper
Baking sheets
Balloons (for U.S. military)
Ball valves (for oil industry)
Birdhouses
Blankets (for fire extinction)
Blast plates
Blocks
Boiler coverings and blankets
Brake blocks
Brake linings
Building board
Buildings ("portable"—that is, prefabricated: cabins, camps, small houses, barracks, hospital isolation wards)
Built-up roof coverings (usually with asphalt)
Bulkheads and marine partitions
Bus shelters (portable)
Bus structure doors

Cable
Cable insulation

Canvas
Capes
Caps
Carburetor-intake attachments (for aircraft)
Carpets
Caulking compound
Ceiling ornaments
Ceiling tile
Cement
Clapboards
Clay
Cleaning compound (for the fur industry)
Cloth
Clutch facings
Coal bins
Cocks (for valves)
Compounded packings
Conduit
Conduit filling
Conveyor belts
Cords
Core plates for drying ovens
Corner beads
Corrugated sheets (for building)
Countertops
Cupboards (over stoves)
Curtain walls

Deck insulation
Deflector plates
Defrosters (for aircraft)
Dental packing
Desks
Diaphragm canopy covers
Dishtowels
Dock and wharf firestops
Doilies
Draperies
Dryers
Duck (fabric)
Ducts
Dummy propellant charges (for ordnance)

Dust-collector bags

85 percent magnesia
Ejectors and lifters (for steam power)
Electric blankets
Electrical cell compartments and doors
Electrical insulation
Electrolytic components
Elevator and dumbwaiter shaft linings
Emulsions
Exhaust covers
Extinguishing agent (for metal powder fires)

Felt
Fencing
Filler (for plastics)
Filling cement
Filters
Finger cots
Firebrick
Fireplaces
Fireplace screens
Firestops
Firewalls
Flange covers
Flanges and fittings (for high-temperature pipes)
Flashing (for roofing)
Floor tile
Flues
Foot powder
Foundry core plates
Friction blocks
Friction tape

Gaiters
Garages (portable)
Gaskets
Gauntlets
Gloves
Gunning mix
Gutters and downspouts

Heat shields
Heater-hose nozzles (for aircraft)

Heating pads
Helmets (for firefighting)
Hoods (for fumes, smoke, and so on)
Hoses

Incandescent thread (for lamps)
Inspirators (for steam power)
Insulation fiber
Insulators (electrical)
Ironing boards and covers
Iron pads, rests, and "slippers"

Joint compound
Joint runners
Journal-box packing

Lab benches
Lace
Lagging
Laminating materials
Lamp parts
Lap
Laundry mangle linings
Laundry tables
Leggings
Linings (for heating and cooling equipment, smokestacks, and so on)
Listings
Locomotive throttle packing
Lumber

Mangle roll covers and pads
Map models
Marine bulkhead panels
Marine valves
Marionette heads
Masonry fill
Mastic
Mattresses (for boilers)
Metal polish
Millboard
Mittens and mitts
Molded friction materials
Molded pipe coverings
Motor and generator windings

Murals

Oven and industrial mitts
Oven linings and insulation

Packing cups
Packings
Pads and mats
Paint and varnish
Pallets
Paneling
Panels and panelboard (electrical)
Paper
Partitions
Patching fiber
Pavement (with asphalt or concrete)
Pipe
Pipe coverings
Plaster
Plastic bonds and fillers
Plugs (electrical)
Potholders
Poultry houses
Prefabricated building panels
Pressure pipe
Projection booths
Proximity suits
Pump valves
Putty

Radiator covers
Railroad enginehouse fire curtains
Railroad passenger-car ceiling liners
Railroad truss plates (including light rail)
Refractories
Refrigerant piping
Refrigerator-car insulation
Rheostat tops
Rollboard
Roofing felt
Roof patching cement
Rope
Rovings
Rugs

School buildings (portable)
Screens
Sculptures
Seals
Sea rings
Seating
Selvedge tape
Septic tank linings
Sewer pipes and joints
Sheetrock
Shelving (for ovens and furnaces)
Shingles
Shovels (for glassmaking)
Shutters
Siding
Sleeves (for proximity clothing)
Sleeving
Slip jackets (for foundries)
Smoke jacks
Smokestack and chimney linings
Socks (for firefighting)
Soffits and fascia
Soldering blocks
Spackle
Spats
Spools
Spray-on insulation
Stage scenery
Stains
Steam hammer packing
Stove board panels
Stove mats
Stove polish
Stucco
Subflooring
Subway sound insulation
Sweating jackets (electrical, used in hospitals)
Switchboards and switch bases

Table mats and pads
Tank linings and covers (for brine, humidifiers, and so on)
Tanks

Tape
Tarpaper
Tent shields (for stoves used in canvas tents)
Theater curtains
Third-rail coverings (for subways)
Thread
Ticket booths
Tile (for ceilings and floors)
Toilet partitions
Transformers
Transmission linings
Trim (architectural)
Tubes (paper, for electric lighting)
Turbine and wrapping jackets
Twine

Underlayment (for flooring)
Upholstery

Valleys (for roofing)
Valve packing
Valves
Varnish
Ventilators
Vent pipe and couplings

Wainscoting
Wallboard
Wallpaper
Washers (for the glass industry)
Waste containers (for hot ashes)
Water pipes and joints
Wicks

Yarn

APPENDIX B. SELECTED LIST OF ORGANIZATIONS SPECIFYING ASBESTOS IN CODES, STANDARDS, OR RECOMMENDATIONS, 1880–1980

Air Conditioning and Refrigerating Machinery Association
Air Filter Institute
American Association of Port Authorities
American Association of University Teachers of Insurance
American Boiler Manufacturers Association and Affiliated Industries
American Electric Railway Association
American Foundrymen's Association
American Gas Association
American Institute of Architects
American Institute of Electrical Engineers
American Marine Insurance Syndicates
American Marine Standards Committee
American Oil Burner Association
American Petroleum Institute
American Railway Engineering and Maintenance of Way Association
American Road Builders Association
American Society for Testing Materials
American Society of Heating and Ventilating Engineers
American Standards Association
American Transit Association
American Water Works Association
Arizona Equitable Rating Office
Arkansas Fire Prevention Bureau
Asphalt Shingle and Roofing Institute
Associated Factory Mutual Fire Insurance Companies
Associated Lumber Mutual Fire Insurance Companies
Association of Canadian Fire Marshals
Association of Edison Illuminating Companies
Association of Fire Underwriters of Baltimore City
Association of Mill and Elevator Mutual Insurance Companies
Association of the Fire Alarm Industries

Board of Fire Underwriters of the Pacific
British Columbia Insurance Underwriters' Association
British Department of Scientific and Industrial Research
British Merchant Shipping Construction Rules
Building Officials Conference of America
Bureau of Explosives

Canadian Automatic Sprinkler Association
Canadian Manufacturers' Association
Canadian Underwriters' Association
Cast Iron Pipe Research Association
Cellulose Plastics Research Association
Central Station Fire Protection Association
Central Traction and Lighting Bureau
Chemical Fire Extinguisher Association
Chicago Board of Underwriters
Chlorine Institute
Cincinnati City Council
City of Boston
City of Dayton, Ohio
City of Detroit
City of Elmira, N.Y.
City of New York
City of St. Louis
City of Spokane
Clay Products Association
Commonwealth of Massachusetts
Commonwealth of Pennsylvania
Compressed Gas Manufacturers' Association
Concrete Institute
Cotton Insurance Association

Diesel Engine Manufacturers' Association

Eastern Underwriters' Inspection Bureau
Edison Electric Institute

Factory Insurance Association
Factory Mutual Engineering Corporation
Factory Mutual Laboratories
Farmers Mutual Insurance Company of Nebraska
Farmers Mutual Reinsurance Association
Federation of Mutual Fire Insurance Companies
Fire Extinguisher Manufacturers Association

Fire Insurance Rating Bureau of Wisconsin
Fire Underwriters' Inspection Bureau

Gypsum Association

Housing and Home Finance Agency

Idaho Surveying and Rating Bureau
Illinois Fire Prevention Association
Illinois Inspection Bureau
Illuminating Engineering Society
Improved Risk Mutuals
Indiana Inspection Bureau
Institute of Makers of Explosives
Institute of Radio Engineers
Insurance Association of Providence
International Association of Electrical Inspectors
International Association of Industrial Accident Boards and Commissions
International Association of Municipal Electricians
International Acetylene Association
International Association of Fire Fighters
International Brotherhood of Electrical Workers
Iowa Insurance Service Bureau
Iowa State College

Kansas Inspection Bureau
Kentucky Actuarial Bureau
Kentucky Inspection Bureau
Heating, Piping and Air Conditioning Contractors National Association

Lightning Rod Manufacturers Association
Louisiana Rating and Fire Prevention Bureau

Manufacturing Chemists' Association of the United States
Michigan Inspection Bureau
Michigan State Association of Mutual Insurance Companies
Middle Department Rating Association
Mississippi State Rating Bureau
Missouri Inspection Bureau
Motion Picture Producers and Distributors of America
Motor Fire Apparatus Manufacturers' Association
Mountain States Inspection Bureau
Mutual Farm Underwriters
Mutual Fire Inspection Bureau

Mutual Fire Insurance Association
Mutual Fire Prevention Bureau

National Association of Air Filter Manufacturers
National Association of Building Owners and Managers
National Association of Credit Men
National Association of Domestic and Farm Pump Manufacturers
National Association of Engine and Boat Manufacturers
National Association of Fan Manufacturers
National Association of Mutual Insurance Companies
National Automatic Sprinkler and Fire Control Association
National Board of Fire Underwriters
National Board of Marine Underwriters
National Clean Up and Paint Up Campaign Bureau
National Electric Light Association
National Electrical Contractors' Association
National Electrical Manufacturers' Association
National Fire Protection Association
National Lumber Manufacturers Association
National Municipal Signal Association
National Paint, Varnish, and Lacquer Association
National Safety Council
National Warm Air Heating and Air Conditioning Association
Nebraska Inspection Bureau
New Brunswick Board of Fire Underwriters
New England Insurance Exchange
Newfoundland Board of Fire Underwriters
New Hampshire Board of Underwriters
New Jersey Schedule Rating Office
New York Board of Fire Underwriters
New York Fire Insurance Rating Organization
Northwestern Association of Electrical Inspectors
Nova Scotia Board of Fire Underwriters

Ohio Farm Bureau Federation
Ohio Inspection Bureau
Oil Insurance Association
Oklahoma Inspection Bureau
Oregon Insurance Rating Bureau
Oregon State Fire Marshal Department

Pacific Coast Building Officials Conference
Pacific Factory Insurance Association

Philadelphia Fire Underwriters' Association
Port of New York Authority

Railroad Insurance Association
Railway Fire Protection Association
Rigid Steel Conduit Association

Safe Manufacturers' National Association
Society of Grain Elevator Superintendents of North America
Society of Motion Picture Engineers
South-Eastern Underwriters' Association

Telephone Group
Tennessee Inspection Bureau
Terminal Elevator Grain Merchants' Association
Texas Inspection Bureau
Texas State Fire Insurance Commission

Underwriters' Association of the District of Columbia
Underwriters' Laboratories
Underwriters' Service Association
U.S. Department of Agriculture, Bureau of Agricultural Economics
U.S. Department of Agriculture, Soils and Agricultural Engineering
U.S. National Bureau of Standards
U.S. Naval Civil Engineering Laboratory
U.S. Waterways Experiment Station

Village of Port Chester, N.Y.
Virginia Insurance Rating Bureau

Washington Surveying and Rating Bureau
Western Actuarial Bureau
Western Canada Insurance Underwriters' Association
Western Factory Insurance Association
Western Sprinkled Risk Association
Western Underwriters' Association
West Virginia Fire Underwriters' Association
West Virginia Inspection Bureau

NOTES

CHAPTER 1 THE ASBESTOS TECHNOLOGY DECISION ENVIRONMENT

1. Sara Wermiel, *The Fireproof Building* (Baltimore: Johns Hopkins University Press, 2000), 140–41, 192–99; see also Jeffrey Karl Ochsner, "Meeting the Danger of Fire: Design and Construction in Seattle after 1889," *Pacific Northwest Quarterly* 93, no. 3 (2002): 115–26; and H. W. Forster, "The Fireproof Building: Its Advantages and Its Weaknesses," *NFPA Quarterly* 7, no. 3 (1914): 403–15. Boston developed America's first attempt at a building code in 1631 after a major fire. On Britain, Europe, and the American colonies, see James C. Robertson, *Introduction to Fire Prevention,* 5th ed. (Upper Saddle River, N.J.: Prentice Hall, 2000), chap. 1. For London's response to the Great Fire of 1666, see James Leasor, *The Plague and the Fire* (New York: Avon, 1961), 183–85.
2. National Fire Data Center, *Fire in the United States*, 11th ed. (Washington, D.C.: U.S. Department of Commerce, Fire Administration, National Fire Data Center, 1999).
3. Denise Guess and William Lutz, *Firestorm at Peshtigo: A Town, Its People, and the Deadliest Fire in American History* (New York: Holt, 2002), 65–125.
4. Grace Stageberg Swenson, *From the Ashes: The Story of the Hinckley Fire of 1894* (St. Cloud, Minn.: North Star, 1979), 4–77.
5. Francis M. Carroll and Franklin R. Raiter, *The Fires of Autumn: The Cloquet–Moose Lake Disaster of 1918* (St. Paul: Minnesota Historical Society, 1990).
6. Anthony Mitchell Sammarco, *The Great Boston Fire of 1872* (Dover, N.H.: Arcadia, 1997); Wermiel, *Fireproof Building*, 5.
7. Nat Brandt and Cathlyn Schallhorn, *Chicago Death Trap: The Iroquois Theatre Fire of 1903* (Carbondale: Southern Illinois University Press, 2003); and Anthony P. Hatch, *Tinder Box: The Iroquois Theatre Disaster, 1903* (Chicago: Academy Publishing, 2003).
8. Mary J. Schneider, *Midwinter Mourning: The Boyertown Opera House Fire* (Boyertown, Pa.: MJS, 1991).
9. Edward T. O'Donnell, *Ship Ablaze: The Tragedy of the Steamboat General Slocum* (New York: Broadway Books, 2003).
10. Gordon Thomas and Max Morgan Witts, *The San Francisco Earthquake* (New York: Stein and Day, 1971).
11. Leon Stein, *The Triangle Fire* (Philadelphia: Lippincott, 1962); and Dave Von Drehle, *Triangle: The Fire That Changed America* (New York: Atlantic Monthly Press, 2003). I refer only to fires here, not to explosions.
12. Bill Minutaglio, *City on Fire: The Forgotten Disaster That Devastated a Town and Ignited a Landmark Legal Battle* (New York: HarperCollins, 2003); and Hugh W. Stephens, *The*

Texas City Disaster, 1947 (Austin: University of Texas Press, 1997); see also Fire Prevention and Engineering Bureau of Texas, *Texas City, Texas, Disaster, April 16, 17, 1947* (Dallas: National Board of Fire Underwriters, 1947).

13. Hal Burton, *The Morro Castle* (New York: Viking, 1973); Gordon Thomas, *Shipwreck: The Strange Fate of the Morro Castle* (New York: Stein and Pay, 1972); and Thomas Gallagher, *Fire at Sea: The Mysterious Tragedy of the Morro Castle* (Guilford, Conn.: Lyons, 2003). On the implications of this disaster for asbestos use as fire prevention in ships, see William Hines, *The Third Wave of Asbestos Disease: Asbestos in Place* (Washington, D.C.: Workplace Health Fund, 1990), 21–23.

14. "Cries of Burning Negroes Heard for Blocks," *Natchez Democrat,* April 24, 1940. There is a fictional account of this event in Leedell W. Neyland, *Unquenchable Black Fires* (Tallahassee, Fla.: Leney Educational, 1994). On the Cocoanut Grove fire, see Robert Selden Moulton, *The Cocoanut Grove Night Club Fire, Boston, November 28, 1942* (Boston: National Fire Protection Association, 1943); Paul Benzaquin, *Fire in Boston's Cocoanut Grove; Holocaust!,* 2d ed. (Boston: Branden, 1967); Edward Keyes, *Cocoanut Grove* (New York: Atheneum, 1984); and Frederica Weeks, "After Cocoanut Grove," *Atlantic Monthly* 171 (1943), 55–57.

15. Stewart O'Nan, *The Circus Fire: A True Story* (New York: Doubleday, 2000); on the legal significance of this fire, see Henry S. Cohn, *The Great Hartford Circus Fire: The Creative Settlement of Mass Disasters* (New Haven, Conn.: Yale University Press, 1992).

16. *Encyclopedia of Cleveland History* (Cleveland, Ohio: Case Western Reserve University, 1998), *http://ech.cwru.edu/ech-cgi/article.pl?id=CSF.*

17. H. W. Forster, *Fire Protection for Schools* (Boston: National Fire Protection Association [hereafter NFPA], prepared and printed for the U.S. Department of the Interior, Bureau of Education, November 1919), 3, 16–17, 24.

18. *Chicago Tribune*, December 2–10, 1958, front section.

19. David Cowan and John Kuenster, *To Sleep with the Angels: The Story of a Fire* (Chicago: Dee, 1996).

20. Professional fire-engineering analysis of this fire appears in Chester I. Babcock and Rexford Wilson, "The Chicago School Fire," *Quarterly of the NFPA* 52, no. 3 (1959): 155–75.

21. The civil defense concerns of American fire professionals early in World War II are set forth in Horatio Bond, *Fire Defense* (Boston: NFPA, 1941).

22. Sir Aylmer Firebrace, "Britain's Wartime Fire Service," and "The Reorganization of the British Fire Service," in *Fire and the Air War,* ed. Horatio Bond (Boston: NFPA, 1946), 23–60. See also Eric L. Bird's article, "Town Planning Lessons of the Fire Blitz" in the same volume.

23. One example is Barbara Nixon, *Raiders Overhead: A Diary of the London Blitz* (London: Scolar/Gulliver, 1980). The first eleven chapters were originally published in 1943.

24. James K. McElroy, "The Work of the Fire Protection Engineers in Planning Fire Attacks," and Robert N. Nathans, "Making the Fires that Beat Japan," in *Fire and the Air War*, 122–48. On American urban insurance maps, see William S. Peterson Hunt and Evelyn L. Woodruff, *Union List of Sanborn Insurance Maps Held by Institutions in the United States and Canada* (Sacramento, Calif.: Western Association of Map Libraries, 1977); and Diane L. Oswald, *Fire Insurance Maps: Their History and Applications* (College Station, Tex.: Lacewing, 1997).

25. Horatio Bond, "The Fire Attacks on German Cities," *NFPA Quarterly* 39 no. 3, (1946): 246; Bond, "Fire Casualties of the German Air Attacks," and Anthony J. Mullaney, "German Fire Departments under Air Attack," in *Fire and the Air War*, 98–121. The illustrations in these articles are not for the faint of heart. There is a more popular account of Hamburg's firestorm in Martin Caidin, *The Night Hamburg Died* (New York: Ballantine, 1979).
26. For example, on Berlin and Cologne, see George N. Shuster, *The Ground I Walked On: Reflections of a College President* (Notre Dame, Ind.: University of Notre Dame Press, 1969), 166, 217.
27. Robert Guillain, *I Saw Tokyo Burning: An Eyewitness Narrative from Pearl Harbor to Hiroshima*, trans. William Byron (London: Murray, 1981), 173–98; see also E. Bartlett Kerr, *Flames over Tokyo: The U.S. Army Air Forces' Incendiary Campaign against Japan, 1944–1945* (New York: Fine, 1991), 139–214.
28. Caidin, *A Torch to the Enemy*, 191.
29. Forrest J. Sanborn, "Fire Protection Lessons of the Japanese Attacks," in *Fire and the Air War*, 169–87.
30. George Wissowa, *Paulys Real-Encyclopädie der Classischen Altertumswissenschaft* (Stuttgart, Germany: Metzlerscher, 1894), series 1830.
31. I present these classical references in their traditional form; readers unfamiliar with Latin and Greek may wish to refer to the Loeb editions of these works. In the case of Gaius Plinius Secundus, this would be D. E. Eichholtz's and Harris Rackham's translation (Cambridge, Mass.: Harvard University Press, 1938).
32. This unaccountable error is propagated in the following sources, among others: Barry I. Castleman and Stephen L. Berger, *Asbestos: Medical and Legal Aspects*, 4th ed. (Englewood Cliffs, N.J.: Aspen, 1996), 1–2, in which it is the opening line of the book; Robert Proctor, *Cancer Wars* (New York: Basic Books, 1995), 111; Landye Bennett Blumstein, LLP, Attorneys, "Ancient History of Asbestos," *http://www.asbestos-attorney-oregon.com/pages/history.html*, accessed on February 9, 2003, which includes several other errors regarding ancient knowledge of asbestos; Michael Bowker, *Fatal Deception* (Emmaus, Pa.: Rodale, 2003); and Roger G. Worthington, P.C., "History of Medical Knowledge," *http://www.mesothel.com/mesothel/pages/medhx_pag.htm*, accessed on January 29, 2003. As recognized by classicists and archaeologists, the authorities on asbestos in the ancient Mediterranean world are Harry Thurston Peck, *Harper's Dictionary of Classical Antiquities* (New York: Harper, 1897); Glorianne Pionati Shams, *Some Minor Textiles in Antiquity* (Gõteborg, Sweden: Åstroms, 1987), 3–11; and James Yates, *Textrinum Antiquorum: An Account of the Art of Weaving among the Ancients* (London: Taylor and Walton, 1843). See also Louise S. Clark, "Textiles and Textile Manufacturing in Ancient Greece: A Survey According to Art-Historical, Archaeological, and Literary Evidence" (master's thesis, University of Wisconsin, 1984), 42. On asbestos in ancient China and Persia, see Bertoldt Laufer, *Sino-Iranica: Chinese Contributions to the History of Civilization in Ancient Iran* (Chicago: Field Museum of Natural History, 1919).
33. Torbern Bergman, *Dissertatio Chemica de Terra Asbestina* (Upsala, Sweden: Edman, 1782); and Clifford Frondel, "Benjamin Franklin's Purse and the Early History of Asbestos in the United States," *Archives of Natural History* 15, no. 3 (1988): 281–87.
34. ASTM E84, *Method of Test of Surface Burning Characteristics of Building Materials.* This standard is also known as NFPA no. 255 and UL 723. In Canada it is incorporated into the Canadian National Building Code as CBD-45, *Flame Spread*; G. W. Shorter, "CBD–45.

Flame Spread," *Canadian Building Digest* (September 1963), *http://irc.nrc-cnrc.gc.ca/cbd/cbd045e.html,* accessed on January 10, 2004. For a history of the flame-spread test and its origin in the 1910s, see A. J. Steiner, "Fire Tests and Fire Protection Engineering," *NFPA Quarterly* 52 no. 3 (1959): 209–20. Its predecessor was ASTM E119.

35. "Asbestos Roof," *Country Life* 44 (1923): 75.

36. NFPA, *Fire Protection Handbook*, 17th ed. (Quincy, Mass.: NFPA, 1991), 6–71; see also David E. Newton, *Encyclopedia of Fire* (Westport, Conn.: Oryx, 2002), 7–8.

37. H. Selleck and A. Whittaker, *Occupational Health in America* (Detroit: Wayne State University Press, 1962), 36; see also "Fire Test of a Theatre Curtain," *Asbestos* 6, no. 1 (1924): 14, 17. On ships, see Frank Rushbrook, *Fire Aboard: The Problems of Prevention and Control in Ships and Port Installations* (New York: Simmons-Boardman, 1961), 257, 330–47; and A.J.S. Bennett, *Ship Fire Prevention* (Oxford: Pergamon, 1964), 2–3.

38. Sammarco, *Great Boston Fire of 1872*, 2, 79, 95; "Unsatisfactory Sprinkler Performance by Occupancy," *Crosby-Fiske-Forster NFPA Handbook of Fire Protection*, 10th ed. (Boston: NFPA, 1948), 914–18; "Elements of Fire-Resistive Construction," *Crosby-Fiske-Forster NFPA Handbook of Fire Protection*, 8th ed. (Boston: NFPA, 1935), 336–50; Walter C. Voss, *Fireproof Construction* (Toronto: Van Nostrand Reinhold, 1948), 267. See also *Fire Protection Engineering: A Symposium of Papers Presented at a Summer Conference at the Massachusetts Institute of Technology, Cambridge, Massachusetts, June, 1942* (Boston: NFPA, 1943).

39. National Board of Fire Underwriters, *Building Code Recommended by the National Board of Fire Underwriters,* 5th ed. (New York: National Board of Fire Underwriters, 1934), 116–17, 161, 168, 181–83, 193–201, 275–79; 1943 edition, 173–255, 290–91, 339–51; 1955 golden anniversary edition, 33–37, 163, 214, 226–55; "Fire Resistance Ratings," 1–43; Factory Mutual Engineering Corporation, *Handbook of Industrial Loss Protection*, 2d ed. (New York: McGraw-Hill, 1967); *Uniform Building Code*, 1967 ed. (Pasadena, Calif.: International Conference of Building Officials, 1967), code standards 6-1-67, 32-9-67, 42-1-67, 43-1-67, 48-1-67, 48-2-67; and "Uniform Mechanical Code," 2, 38–111; 1988 ed., 614–18. Some prefer to reserve the word *code* for bodies of municipal law; but the titles of these sets of guidelines call them codes, and they were enforced by insurance underwriters by denial of coverage.

40. *Crosby-Fiske-Forster NFPA Handbook of Fire Protection*, 8th ed., 374–75.

41. *Building Code of the City of Dayton, Ordinance No. 10383* (Dayton, Ohio: City Commission, 1916), 36, 256; *Building Zone Ordinance: Building Code, Plumbing Code, Village of Port Chester, N.Y.* (Port Chester, N.Y.: Board of Trustees, 1927), 92–95; *Building Code of the District of Columbia Effective July 1, 1930* (Washington, D.C.: U.S. Government Printing Office, 1930): 104–51; *Building Code of the City of New York* (New York: Department of Buildings, 1932), 106, 128–29, 194–95, 250–53, 326; and *Building Code of the City of Cincinnati Effective April 1, 1933* (Cincinnati: City of Cincinnati, 1933), 92–93, 143, 206, 213, 232–33, 246–47, 281–85, 313. For an example of a code that included liability for ignition of neighbors' roofs with flying brands, see Fire Commission of the City of Elmira, N.Y., *Building Code of the City of Elmira, N.Y. Effective July 25, 1923* (Elmira: Fire Commission of the City of Elmira, N.Y., 1923), 43, 54–57, 109, 122–29. Codes of this type are credited with reducing fire loss in pre–World War II Japan; see U.S. Strategic Bombing Survey, Physical Damage Division, *A Report on Physical Damage in Japan* (Washington, D.C.: U.S. Strategic Bombing Survey, 1947), 55.

42. Jacqueline Karnell Corn and Morton Corn, "Changing Approaches to Assessment of Environmental Inhalation Risk: A Case Study," *Milbank Quarterly* 73, no. 1 (1995): 100, 102.

43. Stephen J. Carroll et al., *Asbestos Litigation Costs and Compensation: An Interim Report* (Santa Monica, Calif.: Rand Institute for Civil Justice, 2002). See also Joseph B. Treaster, "Insurer Adds $2 Billion to Asbestos Reserve," *New York Times*, January 15, 2003, pp. C1–2.

44. Deborah Warner, "Roofing, Lagging and Packing: The History of Asbestos Use in the United States" (exhibition at Smithsonian Institution, 1997); and W. Kip Viscusi, *Reforming Products Liability* (Cambridge, Mass.: Harvard University Press, 1991), 18–22.

45. On American health care exceptionalism, see Colin Gordon, *Dead on Arrival: The Politics of Health Care in Twentieth-Century America* (Princeton, N.J.: Princeton University Press, 2003); and Daniel S. Hirshfield, *The Lost Reform: The Campaign for Compulsory Health Insurance in the United States from 1932–1943* (Cambridge, Mass.: Harvard University Press, 1970). On some of the costs to only one of many other nations, see Adam Raphael, *Ultimate Risk: The Inside Story of the Lloyd's Catastrophe* (New York: Four Walls Eight Windows, 1995).

46. There are notorious exceptions, such as Michael Fumento, "The Asbestos Rip-Off," *American Spectator* 22 (October 1989): 21–26.

47. Stephen L. Berger, "Alternatives to Asbestos Insulation," in Castleman and Berger, *Asbestos*, 437–513. On patents, see Julliard C. Lowell, "The Patenting Performance of 1,000 Inventors during Ten Years," *American Journal of Sociology* 37, no. 4 (1932): 569; and Minda Zetlin, "A University Teaches Would-Be Inventors to Ply Their Trade," *Management Review* 78, no. 2 (1989): 42.

48. Morton Sontheimer, "Our National Gamble: Firetrap Schools," *Parents' Magazine* 37 (January 1952): 30–32, 66–67.

49. U.S. Centers for Disease Control [hereafter CDC], National Center for Health Statistics [hereafter NCHS], National Vital Statistics System, "Table 49. Deaths from Selected Occupational Diseases for Persons 15 Years of Age and Over: United States, Selected Years 1980–99," in *Healthy United States, 2003* (Washington, D.C.: CDC, 2004), 193; and National Institute for Occupational Safety and Health, Division of Respiratory Disease Studies, "Section 7: Malignant Mesothelioma," in *Work-Related Lung Disease Surveillance Report* (Cincinnati, Ohio: U.S. Department of Health and Human Services, CDC, Public Health Service, 2002), 164. The CDC changed its method of counting and coding these deaths in 1999 so that more different types of malignant pleural neoplasms were included. This had the effect of increasing the reported death toll from 1,069 in 1998 to 2,485 in 1999, but, of course, the two figures are not comparable. Like fire death statistics, these figures conform to Maines's Law of Research: No data are ever reported in the desired format.

50. A.I.G. McLaughlin, "Talc Pneumoconiosis," *British Journal of Industrial Medicine* 6 (1949): 184–93.

51. NFPA, *Fire Protection Handbook*, 17th ed., 1–4, table 1-1B.

52. William W. Lowrance, *Of Acceptable Risk: Science and the Determination of Safety* (Los Altos, Calif.: Kaufmann, 1976), 16; see also 23 and 29.

53. A. Tversky and D. Kahneman, "The Framing of Decisions and the Psychology of Choice," *Science* 211 (1981): 453–58. See also P. Slovic and Barry Fischoff, "Response Mode, Framing, and Information-Processing Effects in Risk Assessment," in *New Directions for Methodology of Social and Behavioral Science: The Framing of Questions and the Consistency of Response*, ed. R. M. Hogarth (San Francisco: Jossey-Bass, 1982), 21–36.

54. As the Station (Great White) fire of February 2003 made clear, nightclubs are still fairly fire-prone. See also James R. Gill, Lara B. Goldfeder, and Marina Stajic, "The Happy Land Homicides: 87 Deaths Due to Smoke Inhalation," *Journal of Forensic Sciences* 48, no. 1 (2003): 161–63.
55. London Hazards Centre, "Asbestos Substitutes," in *Asbestos Hazard Handbook*, chap. 4, *http://www.lhc.org.uk*, accessed on October 5, 2003; Brooke T. Mossman, "How Safe Are Asbestos Substitutes?" *Studio Potter* 15 (June 1987): 36–37; and U.S. Department of Health and Human Services, Public Health Service, Agency for Toxic Substances and Disease Registry, "Technical Briefing Paper: Health Effects from Exposure to Fibrous Glass, Rock Wool, or Slag Wool," contract no. 205-1999-00024 (Washington, D.C.: U.S. Department of Health and Human Services, 2002); see also Hines, *Third Wave*, 81–82.
56. Ulrich Beck, *Risk Society: Towards a New Modernity* (London: Sage, 1992), 53.
57. National Research Council [hereafter NRC], Committee on Biologic Effects of Atmospheric Pollutants, *Asbestos: The Need for and Feasibility of Air Pollution Controls* (Washington, D.C.: National Academy of Sciences, 1971), 18–31.
58. Charles Perrow, *Normal Accidents: Living with High-Risk Technologies* (Princeton, N.J.: Princeton University Press, 1999).
59. Niklas Luhmann, *Risk: A Sociological Theory* (New York: de Gruyter, 1993), i.
60. I refer here to the large number of articles published between 1927 and 1964 on the dangers of asbestos to workers, to which bibliographic access was available in hundreds of American libraries. I shall have more to say of the contradictions associated with the "conspiracy of silence" argument in chapter 6.
61. Isabel Allende, *The Infinite Plan: A Novel* (New York: HarperCollins, 1993), 300.
62. See, for example, "Test of Fireproof Airplane and Equipment," *Asbestos* 2, no. 2 (1921): 13–14.
63. Asbestos had been known for many centuries but was not widely used before 1900. See, for example, the advertisement for "H. W. Johns Asbestos Plastic Stove-Lining," 1887, reprinted in *Those Were the Good Old Days*, ed. Edgar Jones (New York: Fireside Books of Simon and Schuster, 1959), 36.
64. U.S. Department of Commerce, Bureau of the Census, *Census of Manufactures: Industry Statistics* (Washington, D.C.: U.S. Government Printing Office, 1947, 1954, 1958).
65. K. C. Barnaby, *Some Ship Disasters and Their Causes* (London: Hutchinson, 1968), 227–49, 253–54.
66. Alfred Cecil Hardy and Edward Tyrrell, *Shipbuilding: Background to a Great Industry* (London: Pitman, 1964), 66–72; and P. S. Thorsen and Company, "Insulation Plays an Important Part in the Economy of Ship Operation," in *Shipbuilding in the United States*, published by the National Council of American Shipbuilders on the occasion of the International Meeting of Naval Architects and Marine Engineers, held under the auspices of the Society of Naval Architects and Marine Engineers, New York City, September 14–19, 1936.
67. "85% Magnesia and its Interesting Story," *Asbestos Worker* 5, no. 8 (1926); and 6, no. 1 (1927), cited in David Ozonoff, "Failed Warnings: Asbestos-Related Disease and Industrial Medicine," in *The Health and Safety of Workers*, ed. R. Bayer (New York: Oxford University Press, 1988), 144. See also Armstrong Cork Company, *Nonpareil High Pressure Coverings, for High Pressure and Superheated Steam Lines, Boilers, Breechings and All Heated Surfaces . . . Armstrong Cork Company . . . Pittsburgh, Pa.* (Cleveland, Ohio: Corday and Gross, 1911), 7–8.

68. Jules Backman et al., *War and Defense Economics* (New York: Rinehart, 1952), 118–43; Oliver Bowles, "Asbestos: A Materials Survey," information circular no. 7880 (Washington, D.C.: U.S. Bureau of Mines, 1959); D. V. Rosato, *Asbestos: Its Industrial Applications* (New York: Reinhold, 1959); U.S. Congress, *U.S. Code*, title 32, "National Defense," part 1172, "Asbestos Textiles; Conservation Order M-128 As Amended July 4, 1942," "Conservation Order M-123 As Amended December 14, 1942"; part 1064, "Asbestos; Conservation Order M-79 As of September 30, 1942," "Conservation Order M-283 As Issued October 1943"; and part 3301, "Cork, Asbestos and Fibrous Glass, Conservation Order M-283 As Amended August 11, 1945"; U.S. Department of Interior, U.S. Bureau of Mines for the Materials Office, National Security Resources Board, *1950 Materials Survey: Asbestos* (Washington, D.C.: U.S. Government Printing Office, February 1952); U.S. Department of Labor, Bureau of Labor Statistics, "Asbestos Workers," in *Occupational Outlook Handbook: Employment Information on Major Occupations for Use in Guidance*, bulletin no. 998 (Washington, D.C.: U.S. Government Printing Office, 1951), 387–88; U.S. Department of War, *Strategic and Critical Materials* (Washington, D.C.: U.S. Government Printing Office, 1942); U.S. Office of the President, Office of Emergency Planning, *Stockpile Reports to the Congress* (Washington, D.C.: Office of Defense Mobilization, 1956–62); U.S. Federal Trade Commission, *Gypsum and Asbestos Products and Roof Coating (Except Paint) Manufacturing Corporations* (Washington, D.C.: U.S. Government Printing Office, March 25, 1941); and U.S. Tariff Commission, *Asbestos: A Summary of Information on Uses, Production, Trade and Supply*, industrial materials series report no. M-3 (Washington, D.C.: U.S. Government Printing Office, December 1951).
69. U.S. Congress, House of Representatives, Committee on Education and Labor, Subcommittee on Labor Standards, *Oversight Hearing*, 97th Congress, 2d sess. (1982), 9; and *Hearings on Compensation for Occupational Diseases*, 99th Congress, 1st sess. (1985), 337–38.
70. U.S. Bureau of Mines, *Minerals Yearbook* (Washington, DC: U.S. Government Printing Office, 1947, 1952, 1960, 1967, 1970, 1976).
71. U.S. Congress, House of Representatives, Committee on Education and Labor, Subcommittee on Labor Standards, *Occupational Health Hazards Compensation Act of 1982*, hearings, 97th Congress, 2d sess. (1982), 249.
72. W. Keith C. Morgan and J. Bernard L. Gee, "Asbestos-Related Diseases," in *Occupational Lung Diseases*, ed. W. Keith C. Morgan and Anthony Seaton (Philadelphia: Saunders, 1994), 308–73.
73. Samuel Miller Brownell, "School Buildings Should Be Safe," *School Life* 36 (May 1954), inside cover; N. L. Engelhardt, Sr., "Safeguarding Schools against Fire Dangers," *American School Board Journal* 128, no. 2 (1954): 65–67; Frank G. Lopes, "School Design in 1952," *Architectural Record* 111 (January 1952): 138–41; Helen K. Mackintosh, "Are Today's Children Safe from Fire?" *School Life* 33 no. 5 (1951): 74–76; Albert Q. Maisel, "How Safe Is Your Child's School?" *Woman's Home Companion* 80 (November 1953): 46–71; Morton Sontheimer, "Our National Gamble: Firetrap Schools," *Parents' Magazine* 37 (January 1952): 30–32, 66–67; and N. E. Viles, *School Fire Safety*, bulletin 1951, no. 13 (Washington, D.C.: Office of Education, Federal Security Agency, 1951).
74. See, for example, "Who's Got the Bag?" *Business Week*, May 12, 1956, for an account of one American manufacturer's desperate (and unsuccessful) efforts to obtain asbestos from the Soviet Union.
75. Architects were expected to design to the standards of their time and, in a case in the late 1980s, were held not to be responsible for meeting standards developed several

decades later. Norman Coplan, "Law: Standards of Care," *Progressive Architecture* 71, no. 5 (1990): 59.

76. U.S. Department of Commerce, Bureau of the Census, *Historical Statistics of the United States: Colonial Times to 1970* (Washington, D.C.: U.S. Government Printing Office, 1976): part 1, 143.

77. U.S. Department of Health, Education, and Welfare, Public Health Service, *Vital Statistics of the United States,* vol. 2, *Mortality* (Washington, D.C.: U.S. Government Printing Office, 1932, 1937, 1942, 1949, 1952, 1954, 1957, 1959, 1961, 1966, 1970, 1975).

78. CDC, NCHS, *National Vital Statistics Report* 50, no. 15 (2002): 35.

79. CDC, NCHS, *National Vital Statistics Reports* 52, no. 3 (2003): 1; and Michael J. Karter, Jr., *Fire Loss in the United States during 2001* (Quincy, Mass.: NFPA, Fire Analysis and Research Division, 2002). For historical perspectives on the development of these differences, see Geraldine Fristrom, *Fire Deaths in the United States: Review of Data Sources and Range of Estimates* (Washington, D.C.: U.S. Department of Commerce, National Fire Prevention and Control Administration, National Fire Data Center, 1977). See also CDC, NCHS, *Data Definitions* (Washington, D.C.: CDC, 2003).

80. NFPA, *Fire Protection Handbook*, 17th ed., 1–4.

81. Kearney Management Consultants, *Review of Asbestos Use in Consumer Products* (Washington, D.C.: U.S. Consumer Product Safety Commission, April 1978).

82. Craig Calhoun and Henryk Hiller, "Coping with Insidious Injuries: The Case of Johns-Manville Corporation and Asbestos Exposure," *Social Problems* 35, no. 2 (1988): 162–81.

83. For examples, see State of Alabama, Department of Industrial Relations, Division of Safety and Inspection, *Annual Report* (Montgomery: State of Alabama, 1959); Philip Burch, Jr., *Industrial Safety Legislation in New Jersey* (New Brunswick, N.J.: Rutgers University Press, 1960); California State Legislature, Assembly, Interim Committee on Industrial Relations, Subcommittee on Industrial Safety, *Final Report* (Sacramento: State of California, 1963); New York State Department of Labor, Division of Research and Statistics, *Inspection Activity and Industrial Activity, with Particular Reference to New York State* (New York: State of New York, 1954); and Oklahoma Department of Labor, *Bureau of Factory Inspection: Containing the Laws Governing the Inspection and Regulation of Factories or Other Places Where Labor Is Employed . . .* (Oklahoma City: State of Oklahoma, 1959).

84. Barry I. Castleman and Grace E. Ziem, "Corporate Influence on Threshold Limit Values," *American Journal of Industrial Medicine* 13 (1988): 532.

85. J. G. Townsend, "The Government in Industrial Health," *American Journal of Public Health* 40 (May 1950): 585–87.

86. This device has remained in use for at least half a century; see L. R. Burdick, *A Comparison of the Smokescope and the Ringelmann Chart,* report of investigations no. 5162 (Washington, D.C.: U.S. Bureau of Mines, 1955); U.S. Department of the Interior, U.S. Bureau of Mines, *Ringelmann Smoke Chart (Revision of IC 7718)*, information circular no. 8333 (Washington, D.C.: U.S. Government Printing Office, 1967); "Particulate Emissions," in *Encyclopedia of Environmental Science and Engineering*, 3d ed. (Philadelphia: Gordon and Breach, 1991), 822–32. For the events that in part motivated Pittsburgh's action on air quality, see Lynne Page Snyder, "'The Death-Dealing Smog over Donora, Pennsylvania': Industrial Air Pollution, Public Health Policy, and the Politics of Expertise, 1948–1949," *Environmental History Review* 18 (spring 1994): 117–39.

87. Ozonoff, "Failed Warnings."

88. Both perceptions are necessary. There is, for example, currently no significant doubt that cigarette smoking is a health hazard, but Americans are willing only to restrict smoking in public spaces and not to make the act of smoking illegal. On asbestos, see "Asbestos: Regulating by Litigating," *Technology Review* 85 (November–December 1982): 76–77.

89. "Proceedings of a Conference to Determine Whether or Not There Is a Public Health Question in the Manufacture, Distribution, or Use of Tetraethyl Gasoline," *Public Health Bulletin* 158 (1925): 93; "The Use of Tetraethyl Lead Gasoline in Its Relation to Public Health," *Public Health Bulletin* 163 (1926): v–viii; David Rosner and Gerald Markowitz, "'A Gift of God?' The Public Health Controversy over Leaded Gasoline during the 1920's," and William Graebner, "Hegemony through Science: Information Engineering and Lead Toxicology, 1925–1965," both in *Dying for Work*, ed. David Rosner and Gerald Markowitz (Bloomington: Indiana University Press, 1987); and Paul Neal et al., "A Study of the Effect of Lead Arsenate Exposure on Orchardists and Consumers of Sprayed Fruit," *Public Health Bulletin* 267 (1941).

90. Rom Harré, *Great Scientific Experiments* (Oxford: Oxford University Press, 1983), 1–22.

91. Asbestos deaths studied by Hoffman in 1918 and Cooke in 1924 were those of single individuals.

92. W. C. Hueper, memorandum in *Drug Reporter*, January 11, 1961.

93. Irving J. Selikoff, J. Churg, and Charles H. Mann, *Biological Effects of Asbestos* (New York: New York Academy of Sciences, 1965).

94. *Annual Report of the Chief Inspector of Factories for the Year 1898*, and *Annual Report of the Chief Inspector of Factories for the Year 1906* (London: His Majesty's Stationery Office, 1898, 1906).

95. Frederick Hoffman, "Mortality from Respiratory Diseases in Dusty Trades (Inorganic Dusts)," *Bulletin of the U.S. Bureau of Labor Statistics* 231 (Washington, D.C.: U.S. Government Printing Office, 1918); and M. Greenburg, "The Montague Murray Case," *American Journal of Industrial Medicine* 3 (1982): 351–56.

96. W. E. Cooke, "Pulmonary Asbestosis," *British Medical Journal,* December 3, 1927, pp. 1024–25; W. Burton Wood and S. Roodhouse Gloyne, "Pulmonary Asbestosis: A Review of One Hundred Cases," *Lancet,* December 22, 1934, p. 1383. The term *disease paradigm* relates to the perception of disease, not to its incidence. Thus, there must have been many historical cases of obstreperous children and flaccid men before the invention of disease paradigms for hyperactivity and erectile dysfunction in the second half of the last century. These perceptions and definitions can change dramatically over relatively short periods of historical time. In the early nineteenth century, "fever" was a disease paradigm that could reasonably be entered on a death certificate. By the end of the century, it had been demoted to a symptom of a variety of other disease paradigms but was not regarded as a disease in itself.

97. J. Beattie and J. F. Knox, "Studies of Mineral Content and Particle Size Distribution in the Lungs of Asbestos Textile Workers," in *Inhaled Particles and Vapours*, ed. C. N. Davies (New York: Symposium Publications of Pergamon Press, 1961), 421; and *Industrial Arts Index* (New York: Wilson, 1930).

98. Phillip Ellman, "Pulmonary Asbestosis: Its Clinical, Radiological, and Pathological Features, and Associated Risk of Tuberculosis Infection," *Journal of Industrial Hygiene* 15 (1933): 181. See also W. B. Wood and S. R. Gloyne, "Pulmonary Asbestosis: A Review of One Hundred Cases," *Lancet* 2 (1934): 1383–1385; and E.R.A. Merewether, "The

Occurrence of Pulmonary Fibrosis and Other Pulmonary Affections in Asbestos Workers," *Journal of Industrial Hygiene* 12 (1930): 253.

99. Robert M. Woodbury, *Workers' Health and Safety: A Statistical Program* (New York: Macmillan, 1927), 139; see also Hoffman, "Mortality from Respiratory Diseases in Dusty Trades"; J. Donnelly, "Pulmonary Asbestos," *American Journal of Public Health* (1933): 1275–81; and Thomas Dormandy, *The White Death: A History of Tuberculosis* (New York: New York University Press, 2000): 43, 74, 82.

100. Philip E. Enterline, "Changing Attitudes and Opinions Regarding Asbestos and Cancer, 1934–1965," *American Journal of Industrial Medicine* 20, no. 5 (1991): 685.

101. "Talk of the Town," *New Yorker*, December 11, 1943.

CHAPTER 2 ASBESTOS BEFORE 1880

1. G. Gjessing, "Socio-Archaeology," *Current Anthropology* 16, no. 3 (1975): 330; Marija Gimbutas, "European Prehistory: Neolithic to the Iron Age," *Biennial Review of Anthropology* 3 (1963): 82; Ernst Manker, "Swedish Contributions to Lapp Ethnography," *Journal of the Anthropological Institute of Great Britain and Ireland* 82, no. 1 (1952): 39–54; S. B. Luce, "Archaeological News and Discussions," *American Journal of Archaeology* 45, no. 4 (1941): 630; and 47, no. 4 (1943): 488; J. G. Wilson, "The Addition of Talc and Asbestos to Pot Clay by Past and Present Inhabitants of Karamoja District in Uganda and Adjoining Districts of Kenya," *Man*, n.s. 8, no. 2 (1973): 300–302; Keith Nicklin, "The Location of Pottery Manufacture," *Man* 14, no. 3 (1979): 449; and Robert F. Heizer and Franklin Fenenga, "Archaeological Horizons in Central California," *American Anthropologist* 41, no. 3 (1939): 382. See also "The Prehistory of Finland: the Asbestos-Ceramic Groups ca. 2800–1500 BC," *http://www.nba.fi/NATMUS/Julkais/hist/text.htm*, accessed on July 13, 2003.

2. As in Aeschylus, *Prometheus Bound*, line 527; Plato, *Republic*, sec. 389a; Pindar, *Odes*, book 1, poem 4, line 41; Hesiod, *Theogony*, lines 848, 850; Homer, *Iliad*, book 11, lines 50, 497; and *Odyssey*, book 4, line 583; and in the New Testament, Mark 9:43, Matthew 3:12.

3. Marcus Terentius Varro, *De Lingua Latina* [*On the Latin Language*], trans. Roland Kent (Cambridge, Mass.: Harvard University Press, 1938), 1:125. The supposed corruption of frugal Roman virtues by luxurious Greek introductions was a frequent theme of Roman writers.

4. Asbestos Textile Institute, *Handbook of Asbestos Textiles*, 3d ed. (Pompton Lakes, N.J.: Asbestos Textile Institute, 1967), 1–11.

5. C. Z. Carroll-Porczynski, *Asbestos from Rock to Fabric* (Manchester, Eng.: Textile Institute, 1956), 7–60; see also U.S. Tariff Commission, *Asbestos: A Summary of Information on Uses, Production, Trade and Supply*, industrial materials series report no. M-3 (Washington, D.C.: U.S. Government Printing Office, December 1951), 1, 3.

6. Oliver Bowles, *Asbestos* (Washington, D.C.: U.S. Department of the Interior, U.S. Bureau of Mines, 1937), 3; see also F.C.C. Lynch, "Asbestos, a Canadian Specialty," *Mining and Scientific Press*, April 10, 1920, p. 531.

7. Theophrastus, *De Lapidibus*, trans. D. E. Eichholz (Oxford: Clarendon, 1965), 63.

8. Ibid., 99; Theophrastus, *On Stones: Introduction, Greek Text, English Translation, and Commentary*, ed. and trans. Earle R. Caley and John F. C. Richards (Columbus: Ohio State University, 1956): 48, 87; and Robert H. S. Robertson, "'Perlite' and Palygorskite in Theophrastus," *Classical Review*, n.s. 13, no. 2 (1963): 132. On the Thracian location, see

Oliver Davies, *Roman Mines in Europe* (Oxford: Clarendon Press, 1935), 235. The asbestos identification is in Nathaniel Fish Moore, *Ancient Mineralogy* (1834; reprint, New York: Arno, 1978), 153. Cory Zellers concurs with Moore; see "Asbestos at Its Best," *http://www.emporia.edu/earthsci/amber/go336/zeller*, accessed on June 30, 2003.

9. Alexander Wylie, "Asbestos in China," in *Chinese Researches* (Shanghai: n.p., 1897), 141–54.
10. Edward H. Knight, "Crude and Curious Inventions at the Centennial Exhibition," *Atlantic Monthly* 244 (February 1878): 41–42.
11. Luo Guanzhong [ca. 1330–ca. 1400], *San Kuo Chih Yen I* [*Three Kingdoms*], trans. Moss Roberts (Berkeley: University of California Press, 1991). Lionel Giles remarks on this in "Notes on the Nestorian Monument at Sianfu," *Bulletin of the School of Oriental Studies, University of London* 1, no. 3 (1917): 42.
12. Masumi Chikashige, "Ancient Oriental Chemistry and Its Allied Arts," *Man* 20, no. 81 (1920): 163; Ko Hung, *Alchemy, Medicine, Religion in the China of A.D. 320: The Nei p'ien of Ko Hung (Pao-p'u tzu)*, trans. J. R. Ware (New York: Dover, 1966), 40, 149; and Masumi Chikashige, *Alchemy and Other Chemical Achievements of the Ancient Orient: The Civilization of Japan and China in Early times As Seen from the Chemical Point of View* (Tokyo: Rokakuho Uchida, 1936), 33.
13. John W. Evans, "The Identity of the Amiantos or Karystian Stone of the Ancients with Chrysotile," *Mineralogical Magazine* 14 (1906): 143–48.
14. Nicholas Purcell, in his article on Pliny the Elder in the third edition of the *Oxford Classical Dictionary*, says of his subject that "he not infrequently garbles his information through haste or insufficient thought" (Oxford: Oxford University Press, 1996), 1197. On asbestos in particular, see Moore, *Ancient Mineralogy*, 113, where he mentions Pliny's "fables" regarding asbestos, noting that "it is not unusual with him thus to jumble truth and falsehood."
15. Pliny the Elder, *Natural History*, trans. D. E. Eichholz (Cambridge, Mass.: Harvard University Press, 1962): 10:113. Pliny, *Natural History*, trans. Harris Rackham (Cambridge, Mass.: Harvard University Press, 1950), 5:433.
16. Pliny, *Natural History*, trans. Rackham, 10: 113, 283. As of 2003, Pliny is full-text searchable in Latin at the LacusCurtius website: *http://www.ukans.edu/history/index/europe/ancient_rome/L/Roman/Texts/Pliny_the_Elder*.
17. C. Julius Solinus, *C. Iulii Solini Collectanea Rerum Memorabilium*, ed. Theodor Mommsen (Berlin: Weidmann, 1895).
18. Modern textile scholars do not mention slaves in connection with the working of asbestos in antiquity. See Yates, *Textrinum Antiquorum*, part 1, *On the Raw Material Used for Weaving*, 356–65; R. J. Forbes, *Studies in Ancient Technology* (Leiden: Brill, 1856): 4:62–63, 80; and Glorianne Pionati Shams, *Some Minor Textiles in Antiquity* (Gõteborg, Sweden: Åstroms, 1987), 3–11. Some Greek miners and quarry workers may have been members of unfree classes such as prisoners, slaves, and convicted criminals, but we have no direct evidence of this type of worker in asbestos; see Jean Carpenter and Dan Boyd, "Dragon-Houses: Euboia, Attika, Karia," *American Journal of Archaeology* 81, no. 2 (1977): 208.
19. For examples of this controversy, see Robert Plot, "A Discourse Concerning the Incombustible Cloth Above Mentioned; Address't in a Letter to Mr. Arthur Bayly Merchant, and Fellow of the R. Society; and to Mr. Nicholas Waite, Merchant of London," *Philosophical Transactions* [hereafter *PT*] 15 (1685): 1051–62.

20. Adrian Turnèbe, *M. Ter. Varronis De Lingua Latina Libri qui Supersunt: Cum Fragmentis Ejusdem* (Zweibrücken: Ex Typographia Societatis, 1788).

21. Johann Schild, *C. Suetonius Tranquillus: Et in Eum Commentarius* (Leyden: F. Hackii, 1662), book 2.

22. Marcus Zuerius Boxhorn, *Quaestiones Romanes. Quibus Sacri & Profani Ritus, Eorumgue Caussae* [*sic*] *& Origines, Plurima Etiam Antiquitatis Monumenta, Eruuntur & Explicuntur* (Leyden: D. Lopes de Haro, 1637), question 25; and Isaac Casaubon, *C. Suetonius Tranquillus* (Utrecht: Gisberti a Zyll, 1672), 2:186.

23. The shroud fragments found at Herculaneum were reported to the Royal Society in Britain in C. Paderni and R. Watson, "An Account of the Late Discovories [*sic*] of Antiquities at Herculaneum, &c. in Two Letters from Camillo Paderni, Keeper of the Musaeum Herculanei, to Thomas Hollis, Esq; Translated from the Italian by Robert Watson, M.D., F.R.S.," *PT* 49 (1755): 500.

24. J. G. Keyssler, *Travels through Germany, Bohemia, Hungary, Switzerland, Italy, and Lorrain* (London: Keith, 1760), 2:292; and J. E. Smith, *A Sketch of a Tour on the Continent, in the Years 1786 and 1787* (London: Davis, 1793), 2:201. The Porta cloth was held in the Vatican collections in the late eighteenth century. Other asbestos grave goods were found in Rome in the twentieth century; see Edward A. Heffner, "Archaeological News," *American Journal of Archaeology* 32, no. 3 (July–September 1928): 397, and A. W. van Buren, "News Letter from Rome," *American Journal of Archaeology* 62, no. 4 (1958): 416. On the experiments of the seventeenth and eighteenth centuries, see Edward Lloyd, "An Account of a Sort of Paper Made of Linum Asbestinum Found in Wales in a Letter to the Publisher, from Edward Lloyd of Jesus Coll., Oxon," *PT* 14 (1684): 823–24; and Mr. Wilson [no given name in original], "A Letter from Mr. Wilson to the Publisher, Giving an Account for the Lapis Amianthus, Asbestos, or Linum Incombustible, Lately Found in Scotland," *PT* 22 (1700–1701): 1004–6.

25. Isidore of Charax, *Parthian Stations: An Account of the Overland Trade Route between the Levant and India in the First Century* B.C., trans. Wilfred H. Schoff (Chicago: Ares, 1989), 42, where Schoff quotes *Hou-han-shu*, chap. 88, written in the fifth century A.D. about the first two centuries A.D. See also J. Thorley, "The Silk Trade between China and the Roman Empire at Its Height," *Greece and Rome*, 2d ser. 18, no. 1 (1971): 76, who cites both the *Hou-han-shu* and the *Wei-lio* (A.D. 220–264) in connection with this trade.

26. S. B. Lal, "The Casting of Shellac in Ancient India," *Current Anthropology* 20, no. 2 (1979): 400.

27. Dio Chrysostum (first century A.D.) tells us in his seventy-ninth oration, paragraph 2, that in Roman Imperial times only marble was quarried at Karystos; *Dio Chrysostom*, trans. H. Lamar Crosby (Cambridge, Mass.: Harvard University Press, 1951), 5:307.

28. The *Sung-shou* material is cited in Allen H. Godbey, "The Unicorn in the New Testament," *American Journal of Semitic Languages and Literatures* 56, no. 3 (1939): 264. No asbestos sources west of Syria are mentioned in the ancient Chinese literature.

29. Saint Augustine, *The City of God*, trans. Marcus Dods (New York: Modern Library, 1950), 770–73.

30. Except as noted, the Latin texts of patristic writings cited in this section are from the *Patrologia Latina* [hereafter *PL*], ed. Jacques-Paul Mìgne (Paris: Garnieri, 1844–91), 221 v. in 223. Augustine, Isidore, and some of the others cited here are also available in translation.

31. Eugyppius, *Thesaurus*, in *PL*, 62:786; Bede, *Excerptiones*, in *PL*, 94:551; Rabanus Maurus, *Excerptio de Arte Grammatica Prisciani*, in *PL*, 111:335, 463; Prudentius Trecensis, *De*

Praedestinatione, in *PL*, 115:1336; Isidore of Seville, *Etymologiae*, in *PL*, 82:564; and Marbode of Rennes, *De Lapidibus* (Wiesbaden, Germany: Steiner, 1977), 72, 113. See also J. Horace Nunemaker, "A Comparison of the Lapidary of Marbode with a Spanish Fifteenth-Century Adaptation," *Speculum* 13, no. 1 (1938): 66–67.

32. Albertus Magnus, *Book of Minerals*, trans. Dorothy Wyckoff (Oxford: Clarendon, 1967), 69–70.

33. Eugenius III Toletanus, *Epitaphium*, in *PL*, 87:29; Erigena, *De Divisione Naturae*, in *PL*, 87:960; and Petrus Damianus, *De Divina Omnipotentia in Reparatione Corruptae et Factis Infectis Reddendis*, in *PL*, 145:613.

34. Gaudentius, *Sermones*, in *PL*, 20:959; and Stephanus of Byzantium, *Ethnicorum quae Supersunt* (Berlin: Reimer, 1849), 184–85.

35. Simeone Maiolo, *Dierum Canicularium Tomi VII*, (Bayern: Schönwetteri, 1691), colloquia 18, 22; and Claude Saumaise, *Plinianae Exercitationes in Caji Julii Solini Polyhistora: Item Caji Julii Solini Polyhistor* (Utrecht: van de Water, 1689), chap. 12.

36. Clare Browne, "Salamander's Wool: The Historical Evidence for Textiles Woven with Asbestos Fibre," *Textile History* 34, no. 1 (2003): 66. Browne also comments on the apparent ignorance of western medieval sources, particularly the patristic writers.

37. Faustinus Arevalus, *Isidoriana*, in *PL*, 81:138, 82:990; Jerome (Hieronymus Stridonensis), *Commentaria in Ezechielem*, in *PL*, 25:400; Dracontius, *Carmen de Deo*, in *PL*, 60:851; Rabanus Maurus, *Commentaria in Exodum*, in *PL*, 108:179; Anastasius, *Historia de Vitis Pontificum Romanorum*, in *PL*, 127:1531, 1539; and Godefridus Ghiselbertus, *De Vita Auctoris*, in *PL*, 198:62.

38. W. E. Sinclair, *Asbestos, Its Origin, Production, and Utilization* (London: Mining Publications, 1959), 2.

39. No such miraculous napery is mentioned in either of the two biographies of Charlemagne written immediately after his death; the legend is likely to have been a later invention, dating from the period after the ninth century in which the Frankish king's reputation was gradually inflated to nearly Messianic stature. For an example of an earlier biography, see A. J. Grant, *Early Lives of Charlemagne by Eginhard & the Monk of St. Gall* (London: Chatto and Windus, 1922). The legend is persistent; see Keasbey & Mattison Company, *Legends of Asbestos* (Ambler, Pa.: Keasbey & Mattison Company, 1940), 8; and Becker & Haag Berlin, *Asbestos; Its Sources, Extraction, Preparation, Manufacture and Uses in Industry and Engineering* (Berlin: Becker & Haag Berlin, 1928), 15.

40. Ahmad ibn Muhammad Ibn al-Faq ih al-Hamadh an I, *Abrégé du livre des pays*, trans. Henry Massé (Damascus: Institut Français de Damas, 1973), 251. The original work in Arabic dates from 903.

41. Berthold Laufer cites the *Wei Su*, chap. 102, sec. 4b, in *Sino-Iranica; Chinese Contributions to the History of Civilization in Ancient Iran, with Special Reference to the History of Cultivated Plants and Products* (Chicago: Field Museum of Natural History, 1919), 498. A fictional account of a similar asbestos robe appears in Edison Marshall, *Caravan to Xanadu: A Novel of Marco Polo* (New York: Farrar, Straus, and Young, 1953); see also B. J. Whiting's remarks on this text in "Thirteen Historical Novels," *Speculum* 32, no. 1 (1957): 135.

42. Yates, *Textrinum Antiquorum*, 362–65. The illumination from the *Codex* is reproduced in Herbert Bloch, *Monte Cassino in the Middle Ages* (Cambridge, Mass.: Harvard University Press, 1986), 3:1169. The descriptive text of the "relic" appears in Leo Marsicanus, *Die Chronik von Montecassino = Chronica Monasterii Casinensis* (Hannover,

Germany: Hahn, 1980), 229–30. The passage in question, book 2, chap. 33, is thought to date from late in the year 1021. Further light is thrown on this document by Bernard de Montfauçon (1655–1741), *The Travels of the Learned Father Montfauçon from Paris thro' Italy . . . Made English from the Paris Edition* (London: D. L., 1712), 380–82.

43. Marco Polo, *The Travels of Marco Polo, the Venetian. The Translation of William Marsden Revised,* ed. Thomas Wright (London: Bohn, 1854), 112–13.

44. John Speed, *A Newe Mape of Tartary* (London: Humble, 1626).

45. C. Ricchieri and Lodovicus Caelius Rhodiginus, *Ludovici Caelii Rhodigini Lectionum Antiquarum Libri Triginta* (Frankfurt am Main: Marnium und Aubrium, 1599), book 18, chap. 31.

46. Archibald Rose, "Chinese Frontiers of India," *Geographical Journal* 39, no. 3 (1912): 201–2. Rose apparently does not think to inquire why the chieftain values this garment, which would have been heavy, scratchy, and awkward to wear.

47. Berthold Laufer, *Asbestos and Salamander, an Essay in Chinese and Hellenistic Folk-lore* (Leiden: Brill, 1915). Prester John's letter is reproduced in Sabine Baring-Gould, *Curious Myths of the Middle Ages* (Boston: Roberts, 1867), chap. 2. See also "The Salamander," *Harper's* 5, no. 30 (1852): 764.

48. David M. Sever, *Reproductive Biology and Phylogeny of Urodela* (Enfield, N.H.: Science Publishers, 2003), 5–7; on the ability of at least some salamanders to pass unscathed through fire, see Mark R. Stromberg, "*Taricha Torosa* (California Newt): Response to Fire," *Herpetological Review* 28, no. 2 (1997): 82–84.

49. All living things on earth are hydrocarbons; life is a process of slow combustion. Stephen Pyne, *Fire: A Brief History* (Seattle: University of Washington Press, 2001), 11. Asbestos remained the only economically important mineral fiber until the twentieth century. See, for example, William I. Hannan, *The Textile Fibres of Commerce. A Handbook on the Occurrence, Distribution, Preparation, and Uses of the Animal, Vegetable, and Mineral Fibres Used in Cotton, Woollen, Paper, Silk, Brush and Hat Manufacture* (London: Griffin, 1902), 3, 227, 228.

50 Charles Bonnet, *The Contemplation of Nature* (London: Longman, Becket, and de Hondt, 1766); cited in Stephen Jay Gould, "Bound by the Great Chain," in *The Flamingo's Smile* (New York: Norton, 1985), 285.

51. Aaron J. Ihde, *The Development of Modern Chemistry* (1970; reprint, New York: Dover, 1984), 3–31.

52. Pietro Andrea Mattioli, *Commentaires de M. P. André Matthiolus, médecin senois, sur les six livres de Pedacius Dioscoride, anazarbeen de la matière médicinale* (Lyon: Cotier, 1572): 538–39.

53. Georgius Agricola, *De Natura Fossilium* [*Textbook of Mineralogy*] (New York: Geological Society of America, 1955): 5:93–94. Ole Worm, *Museum Wormiamum* (Leiden: Elsevir, 1655), 55–56; and Athanasius Kircher, *Mundus Subterraneus* (Amsterdam: Jansson at Waesberg and Sons, 1678), book 8, chap. 2.

54. François Rabelais, *The Complete Works of François Rabelais,* trans. Donald M. Frame (Berkeley: University of California Press, 1991), 409–12; Victor Hugo, *Hans of Iceland: Or the Demon of the North, a Romance,* trans. Abby L. Alger (Boston: Estes and Lauriat, 1891), 482; Metro-Goldwyn-Meyer Studios, *The Wizard of Oz* [motion picture] (1939). François Rigolot thinks most of Rabelais's information is drawn from Pliny; see Rigolot, "Rabelais's Laurel for Glory: A Further Study of the 'Pantagruelion,'" *Renaissance Quarterly* 42, no. 1 (1989): 62.

55. Worm, "De Amiantho, Selenitide, Talco, &c.," in *Museum Wormianum,* 55–56; and Kircher, *Mundus Subterraneus,* book 8, chap. 2

56. J. E. Smith, *A Sketch of a Tour on the Continent, in the Years 1786 and 1787* (London: Davis, 1793), 2:201.

57. Matthias Tiling, "Observatio LXI: De Lino Vivo aut Asbestino & Incombustibili," in *Miscellanea Curiosa; Sive, Ephemeridum Medico-Physicarum Germanicarum,* ed. Deutsche Akademie der Naturforscher (Nuremberg: Streck, 1684), 1:109–23.

58. Sinclair, *Asbestos,* 3; John Ciampini, "An Abstract of a Letter, Wrote Some Time Since, by Signior John Ciampini of Rome, to Father Bernard Joseph a Jesu Maria, Etc. Concerning the Asbestus, and Manner of Spinning and Making an Incombustible Cloath Thereof," *PT* 22 (1700): 911–13. See also F. E. Brückmann, *Historia Naturalis Curiosa Lapidis . . . Ejusque Praeparatorum, Chartae Nempe, Lini, Lintei et Ellychniorum Incombustibilium* (Brunswick, 1727), 31–32; and Edward Lloyd, "An Account of a Sort of Paper Made of Linum Asbestinum Found in Wales, in a Letter to the Publisher," *PT* 14 (1684): 823–24. See also Erich Pontoppidan, *The Natural History of Norway* (London: Linde, 1755).

59. Browne, "Salamander's Wool," 68.

60. Saint Augustine and Juan Luis Vives, *Saint Augustine, Of the Citie of God with the Learned Comments of Io. Lodouicus Viues. Englished first by I. H. And now in this Second Edition Compared with the Latine Originall, and in Very Many Places Corrected and Amended* (London: Eld and Flesher, 1620), 790–91; Pietro Della Valle, *The Travels of Sig. Pietro della Valle, a Noble Roman, into East-India and Arabia Deserta* (London: Macock, 1665), 190; Joseph Pitton de Tournefort, Tournefort, *A Voyage into the Levant,* trans. John Ozell (London: Browne, 1718), 1:129; and Charles Sigisbert Sonnini, *Travels in Greece and Turkey* 1 (London: Longman and Rees, 1801), 1:66. See also accounts of the mineral in cabinets of curiosities in Richard Lassels, *An Italian Voyage* (London: Wellington, 1698), 85; John H. Appleby, "Robert Dingley, F.R.S. (1710–1781), Merchant, Architect and Pioneering Philanthropist," *Notes and Records of the Royal Society of London* 45, no. 2 (1991): 141; and Henry Marczali and Stephen Szechenyi, "A Hungarian Magnate at Cambridge in 1787," *Cambridge Historical Journal* 3, no. 2 (1930): 215. The collectible character of asbestos in the seventeenth and eighteenth centuries is noted in Allen Kurzweil's novel, *A Case of Curiosities* (San Diego: Harvest, 1992), 294. Other travelers' accounts include John Ray and F. Willughby, *Observations Topographical, Moral, & Physiological; Made in a Journey Through Part of the Low-Countries, Germany, Italy, and France* (London: Martyn, 1673), 83; and Didier Bugnon, *Rélation exact concernant les caravanes ou cortèges des marchands d'Asie* (Nancy, France: Charlot and Deschamps, 1707), 37–39.

61. "Inquiries for Turky," *PT* 1 (1665): 361; and "A Curious Relation, Taken out the Third Venetian Journal De Letterati, of March 15, 1671; of a Substance Found in Great Quantities in Some Mines of Italy," *PT* 6 (1671): 2167–69.

62. Plot, "A Discourse," 1051–62; Patrick Blair, "Part of a Letter from Mr. Patrick Blair to Dr. Hans Sloane, Royal Society Secretary, Giving an Account of the Asbestos, or Lapis Amiantus, found in the High-Lands of Scotland," *PT* 27 (1710–12): 434–36; J. Harris and J. T. Desaguliers, "An Account of Some Experiments Tried with Mons. Villette's Burning Concave, in June 1718," *PT* 30 (1719): 976–77; and Mr. Angestein [*sic*], "A Short Account of Some Swedish Minerals, &C. Sent from Mr Angestein, Overseer of the King of Sweden's Mines, to Mr James Petiver, Apothecary, and F.R.S.," *PT* 28 (1713): 223. On burning-mirror experiments, see also "Burning-Mirrors and Sun-Machines," *Manufacturer and Builder* [hereafter *MB*] 2, no. 7 (1870): 1.

63. Torbert Bergman, *Physical and Chemical Essays* (London: Murray, 1784), 196–204; on the poor writing qualities of asbestos paper, see Oliver Bowles, "History of Asbestos Paper," *Chemical and Metallurgical Engineering* 22, no. 5 (1920): 208–9. See also "Funeral Dress of Kings," *Better Homes and Gardens* 12 (April 1934): 87.

64. For examples, see *PT* 22 (1700): 1004–6; 45 (1748): 416–89; 51 (1759): 837–38; 64 (1774): 481–91; 82 (1792): 28–47; 97 (1807): 1–56; 112 (1822): 457–82; 136 (1846): 21–40; and 139 (1849): 349–91. For Davy's use of asbestos, see also "Sir Humphry Davy," *Living Age* 4, no. 34 (1845): 9.

65. This artifact is still in the collections of the British Museum. See Jessie M. Sweet, "Benjamin Franklin's Purse," *Notes and Records of the Royal Society of London* 9, no. 2 (1952): 308–9; and Clifford Frondel, "Benjamin Franklin's Purse and the Early History of Asbestos in the United States," *Archives of Natural History* 15, no. 3 (1988): 281–87. On Hamilton, see E. P. Panagopoulos, "Hamilton's Notes in His Pay Book of the New York State Artillery Company," *American Historical Review* 62, no. 2 (1957): 314.

66. Pehr Kalm, *The America of 1750; Peter Kalm's Travels in North America; The English Version of 1770*, ed. A. Benson (New York: Wilson-Erickson, 1937), 1:157, 2:643. Asbestos was found in a forge in France in 1759, but its probable use there as a thermal insulator was not understood either by the proprietor or the Royal Society fellow who reported it; see Turberville Needham, "An Account of a Late Discovery of Asbestos in France: In a Letter to the Rev. Tho. Birch, D. D. Secretary to the Royal Society," *PT* 51 (1759): 837–38.

67. Johann Reinhold Forster and J. Taylor, *An Introduction to Mineralogy or an Accurate Classification of Fossils and Minerals Viz., Earths, Stones, Salts, Inflammables and Metallic Substances*, 1st ed. (London: Johnson, 1768), 16–17; Giacinto Gimma and Antonio Baldi, *Della storia naturale delle gemme, delle pietre, e di tutti i minerali, ovvero, della fisica sotterranea di D. Giacinto Gimma* (Naples: Muzio, 1730), 381; Parker Cleaveland, "Species 58 Asbestus," in *An Elementary Treatise on Mineralogy and Geology*, 2d ed. (Boston: Cummings and Hilliard, 1822), 325–29; Axel Fredrik Cronstedt, *An Essay Towards a System of Mineralogy*, 2d ed. (London: Dilly, 1788), 1:113–19; Samuel Robinson, *A Catalogue of American Minerals, with Their Localities* (Boston: Cummings and Hilliard, 1825), 141, 306; and Emanuel Mendes da Costa, *A Natural History of Fossils* (London: Davis and Reymers, 1757).

68. "Eternal Lamps," *Living Age*, 39, no. 500 (1853): 763; A. M. Mace, "Vapor Lamp Burners," *SA*, 13, no. 43 (1858): 338; "New Lime Light without Oxygen," *SA* 21, no. 15 (1869): 231; "Incombustible Wicks," *MB* 2, no. 9 (1870): 278; "A New Use for Asbestos," *MB* 9, no. 6 (1877): 127; "New Portable Blow-Pipe," *Scribner's Monthly* 16, no. 5 (1878): 756; and "Incombustible Lamp Wicks," *MB* 12, no. 1 (1880): 5. The Hunterian Museum at the University of Glasgow has examples of asbestos lampwicks; see GLAHM 105812 and 113579, *http://www.huntsearch.gla.ac.uk*, accessed on September 7, 2003. Asbestos wicks were still made by hand in some parts of Asia in the 1870s and in Greenland as late as 1937; see H. Trotter, "On the Geographical Results of the Mission to Kashghar, under Sir T. Douglas Forsyth in 1873–74," *Journal of the Royal Geographical Society of London* 48 (1878): 215; and Herman R. Friis, "Greenland: A Productive Arctic Colony," *Economic Geography* 13, no. 1 (1937): 90.

69. Nathan Rosenberg, "On Technological Expectations," *Economic Journal* 86, no. 343 (1976): 532. He cites Charles M. Coleman, *P.G. and E. of California: The Centennial Story of Pacific Gas & Electric Company, 1852–1952* (New York: McGraw-Hill, 1952), 81. These mantles were still in use in propane lamps in the second half of the twentieth century.

70. Eusebe Salverte, "The Occult Sciences," *Living Age* 5, no. 58 (1845): 581.

71. "Jersey Marble," *SA* 5, no. 35 (1850): 276; "Junction of the Atlantic and the Pacific," *Living Age* 27, no. 337 (1850): 199–200; H. Gerber and R. F. Burton, "Geographical Notes on the Province of Minas Geraes," *Journal of the Royal Geographical Society of London* 44 (1874): 278–79; J. R. Hamilton, "The Natural Wealth of Virginia," *Harper's* 32, no. 187 (1865): 33; "Tin in Missouri," *SA* 17, no. 10 (1867): 147; Winslow Cossoul Watson, *The Military and Civil History of the County of Essex, New York.* (Albany, N.Y.: Munsell, 1869), 415; "Asbestos," *MB* 5, no. 5 (May 1873): 114; "Nyassa-Land and Its Commercial Possibilities," *Science* 14, no. 355 (1889): 344; and Alfred Aylward, "Dutch South Africa: Its Hydrography, Mineral Wealth, and Mercantile Possibilities," *Journal of the American Geographical Society of New York* 15 (1883): 13.

72. Warner, "Roofing, Lagging, and Packing"; and "Asbestos," *SA* 4, no. 25 (1849): 195.

73. "New Application of Gun Cotton and Asbestos," *SA* 4, no. 21 (1849): 163; and "List of Patent Claims: To John Allen of Cincinnati, O.," *SA* 7, no. 16 (1852): 126. A similar use is documented in "New Base for Artificial Teeth," *SA* 17, no. 4 (1857): 54.

74. For examples, see *SA* 14, no. 4 (1858): 31; n.s. 1, no. 10 (1859): 159; and 2, no. 23 (1860): 367; see also *MB* 1, no. 1 (1869): 29; and 4, no. 12 (1872), unpaginated backmatter.

75. See "New Inventions," *SA* 11, no. 27 (1856): 212; "M. de Luca obtains Oxygen Gas," 5, no. 9 (1861): 141; "Notes and Queries," 12, no. 1 (1865): 12; "Asbestos, a Material for Glass Making," 17, no. 24 (1867): 374; "The New Zirconia Light," 21, no. 4 (1869): 51; and "Answers to Correspondents," 21, no. 6 (1869): 91. In *MB* see "Asbestos," 5, no. 4 (1873): 74; "Diamagnetism and Paramagnetism," 4, no. 2 (1872): 40; "Electro-plating by Heat," 7, no. 4 (1875): 85; and "A Novel Way to Fire Blasts," 10, no. 9 (1875): 207.

76. "Experiments on the Influence of Air in Promoting the So-Called Spontaneous Generation," *Living Age* 69, no. 891 (1861): 781; "Spontaneous Generation," *SA* 6, no. 22 (1862): 343; and 6, no. 23 (1862): 363.

77. "Arsenic in Wall Paper," *SA* 13, no. 40 (1858): 320; "Improved Carbureting Apparatus," *SA* 6, no. 18 (1862): 280; "Manganese," *SA* 9, no. 12 (1863): 180; "The Uses of Gun-Cotton," *MB* 1, no.1 (1869): 82; "Asbestos," *MB* 5, no. 4 (1873): 74; and "Filtering Alkaline Solutions," *MB* 6, no. 2 (1874): 47.

78. "Electrical News," *Science* 12, no. 306 (1888): 289; "War-Balloons," *Science* 15, no. 369 (1890): 138; L. C. Bernacchi, "Preliminary Report on the Physical Observations Conducted on the National Antarctic Expedition, from 1902 to 1904," *Geographical Journal* 26, no. 6 (1905): 643; Augustus H. Gill, "Suggestions for the Construction of Chemical Laboratories," *Science* 30, no. 773 (1909): 550; Raymond H. Pond, "Laboratory Table Tops," *Science* 33, no. 849 (1911): 529; "The Gray Herbarium," *Science* 41, no. 1062 (1915): 676; "The Mount Everest Expedition," *Geographical Journal* 60, no. 2 (1922): 141; "Photography on the Hamilton Rice Expedition, 1924–25," *Geographical Journal* 68, no. 2 (1926): 145; and "The Antarctic," *Geographical Journal* 75, no. 3 (1930): 254.

79. Asbestos Packing Company, *Price List* (1885); Fairbanks & Company, *Vulcanized Asbestos Disc* (New York: Fairbanks, 1889); and H. W. Johns Manufacturing Company, *Descriptive Price List: Asbestos* (New York: Johns Company, 1889).

80. "Recent Foreign Inventions," *SA* 9, no. 21 (1854): 163; "Green Ink—George Smillie, New York City," in "List of Patent Claims," *SA* 8, no. 15 (1863); 237; "Incombustible Account Books," *MB* 6, no. 9 (1874): 216; "Asbestos Paper," *World's Work* 11, no. 1 (1875): 140; and "Various Practical Uses of Asbestos," *MB* 8, no. 8 (1876): 176.

81. "List of Patent Claims," *SA* 11, no. 36 (1856): 282; 3, no. 16 (1860); 13, no. 14 (1865): 217; and "Dry Journal-Boxes," *MB*, 7, no. 9 (1875): 195.

82. On asbestos in gas appliances, see "Gas Stoves," *SA* 8, no. 4 (1852), 32; "Cooking by Gas," *SA* 1, no. 18 (1859), 290; "Foreign Summary-News and Markets," *SA* 1, no. 22 (1859): 355; "Amianthus, or Asbestos," *MB* 6, no. 1 (1874): 16; and Harvey Levenstein, "The New England Kitchen and the Origins of Modern American Eating Habits," *American Quarterly* 32, no. 4 (1980): 373. For the perceived vulgarity of gas fireplaces, see "Recent Literature," *Atlantic Monthly* 31, no. 186 (1873): 495; Clarence Cook, "Beds and Tables, Stools and Candlesticks," *Scribner's Monthly* 11, no. 3 (1876): 348; and Richard S. Urhbrock, "Popular Usage of the Terms 'Instinct' and 'Instinctive,'" *Scientific Monthly* 34, no. 6 (1932): 546. On mats, see "Contrivance for Retarding Rapid Ebullition," *SA* 12, no. 21 (1865): 321. On the secular worship of the hearth, see Margaret Hindle Hazen and Robert M. Hazen, *Keepers of the Flame: The Role of Fire in American Culture, 1775–1925* (Princeton, N.J.: Princeton University Press, 1992), 61–64.

83. G. Aldini, "Art of Preparing Asbestos and Rendering It Fit for Spinning and Weaving the Use of the Cloth in Garments Worn by Firemen," *American Journal of Science* 18, 1st ser. (1830): 177–79.

84. On nineteenth-century proximity suits, see "Making Gloves," part of a report on a meeting of the Polytechnic Association of the American Institute, *SA* 11, no. 17 (1864): 260.

85. J. T. Donald, "Canadian Asbestus [*sic*]: Its Occurrence and Uses," *Popular Science Monthly* 36 (1889): 530.

86. "Are Iron Safes Safe?" *SA* 15, no. 4 (1866): 55; Edward Everett Hale, "Life in the Brick Moon," *Atlantic Monthly* 25, no. 148 (1870): 217; and R. D. Blackmore [Richard Doddridge], *Alice Lorraine: A Tale of the South Downs*, serialized in *Littell's Living Age*, 5th ser., 10, no. 1608 (1875): 349–50.

87. "Artificial Asbestos, or Mineral Wool," *MB* 10, no. 1 (1878): 5; and "Incombustible Mineral Wool," 9, no. 2 (1877): 42; see also William Peters, U.S. patent 34,283, "Improvement in Packing for Steam and other Engines," *SA* 6, no. 7 (1862): 109.

88. "Water Glass," *SA* 6, no. 2 (1862): 25. See also John A. Church, "Scientific Miscellany," *Galaxy* 22, no. 4 (1876): 566

89. "Memoranda," *Scribner's Monthly* 4, no.1 (1872), 115; and 14, no. 1 (1872): 130–31. Robert H. Jones mentions the falling price of asbestos in the two decades preceding his article, in "Asbestos and Asbestic: With Some Account of the Recent Discovery of the Latter at Danville, in Lower Canada," *Journal of the Society of Arts* 45 (1897): 549.

90. On the costs of processing the mineral, see "Uses of Asbestos," *Chamber's Journal* 74 (1897): 411.

91. John F. Springer, "Asbestos, Its Production and Industrial Applications," *Cassier's Magazine* 42 (1912): 299–300.

92. "Cincinnati Exposition," *MB* 5, no. 11 (1873): 242. See also "Hoisting Engines," *MB* 6, no. 5 (1874): 1.

93. "The Fair of the American Institute," *MB* 7, no. 12 (1875): 272.

94. In *SA*, for examples of advertising and editorial endorsements, see 18, no. 15 (1868): 240; 18, no. 18 (1868): 279, 288; 19, no. 11 (1868): 176; 19, no. 12 (1868): 183, 187, 192; 20, no. 10 (1869): 156, 160; 20, no. 11 (1869): 172; 20, no. 12 (1869): 188, 192; 20, no. 13 (1869): 204; 20, no. 14 (1869), 220, 224; 20, no. 15 (1869): 236; 20, no. 16 (1869): 252; 20, no. 18 (1869): 284; and 20, no. 19 (1869): 300. In *MB*, see 3, no. 12 (1871): 4; 4, no. 4 (1872): 90;

4, no. 5 (1872): 115; 4, no. 12 (1872): 3; 5, no. 2 (1873): 47; 5, no. 11 (1873): 256; 7, no. 4 (1875): 80; 7, no. 12 (1875): advertising page 1; 8, no. 12 (1876): advertising page 1; 11, no. 5 (1879): 116; and 12, no. 12 (1880): 287. On the growth of Johns's company, see "Evolution of a Job," *Nation's Business* 25 (July 1937): 26; and Edwin J. Becker, "Magic Mineral," *American Mercury* 79 (October 1954): 107–10.

95. "Keeping the Heat In," *SA* 174, no. 3 (1946): 122, 124; see also the controversy over the kind and proper amount of asbestos insulation to use in "Value of Sheet Asbestos on Hot Pipes," *Mechanical Engineering* 42 (January–March 1920): 69 (January), 188–89 (March). P. Nicholls recommended air cell rather than paper for this purpose; see his "Warm-Air Furnace Insulation," *American Society of Heating and Ventilation Engineers Journal* 27 (September 1921): 690–92.

96. "Usefulness of Asbestos," *MB* 11, no. 3 (1879): 68.

97. "The Centennial Exposition," *MB* 8, no. 8 (1876): 176; "Asbestos," *MB* 9, no. 5 (1877): 102–3; see also "Steam Pipe and Boiler Coverings," *MB* 11, no. 12 (1879): 268; and "Asbestos Coverings for Steam Boilers and Pipes," *MB* 12, no. 2 (1880): 31; "Complete Combustion Boiler," *Science* 13, no. 333: 492; and "Scientific Miscellany," *Galaxy* 22, no. 1 (1876): 134.

98. Keasbey & Mattison Company, *Trade List, 85% Magnesia and Other Coverings* (Ambler, Pa: Keasbey & Mattison Company, 1907), 3–4; Asbestos and Magnesia Manufacturing Company, *Ehrets 85% Magnesia Coverings* (Philadelphia: Elliott, 1905), 4–5. See also "Keeping the Heat In," *SA* 174, no. 3 (1946): 122.

99. Robert E. Peary, "Party and Outfit for the Greenland Journey," *Journal of the American Geographical Society of New York* 23 (1891): 262.

100. Johns Company, *Descriptive Price List: Asbestos* (1889), 25; (1896), 31; see also "Uses of Asbestos in Electrical Appliances," *Engineering Magazine* 11 (September 1896): 1125.

101. "36,195—William Peters . . . Improvement in the Manufacture of Fire Brick," *SA* 7, no. 9 (1862): 141; "74,587 Composition for Manufacturing Stone . . . A. Pelletier," *SA* 18, no. 10 (1868): 156; "No. 164,850 Improvement in Composition for Building, by Arza T. Lyon," *MB* 7, no. 9 (1875): 202. On fire brick, see also "A New Industry in Tasmania," *MB* 5, no. 6 (1873): 127. On cement, see "80,084—Roofing Cement—J. A. Moore, Providence, R.I.," *SA* 19, no. 6 (1868): 92.

102. "New Buildings in New York and Vicinity," *MB* 6, no. 2 (1874): 33; "An Absolutely Fire-Proof Building," *MB* 7, no. 11 (1875): 256; and "Asbestos Products," *MB* 12, no. 10 (1880): 227.

103. Sold in forty-eight-inch planks, the half-inch thickness, forty-two inches wide, was forty cents a square foot; a 108-square-foot roll was forty dollars. Keasbey & Mattison Company, *Magnesia Building Lumber, Floors, Shingles, Boards, &C* (Ambler, Pa: Keasbey & Mattison Company, 1902), n.p.; and Johns-Manville Company, *J-M Asbestos Roofing; "Look for the White Top," Catalog No. 303* (New York: Johns-Manville Company, 1909), 3. This catalog featured pictures of the material in fire tests, pp. 10–11. On asbestos-cement shingles, see Asbestos Shingles, Slate & Sheathing Company, *Trade Priced List of the Asbestos Roofing Slates, Shingles and Sheathings* (Ambler, Pa: Asbestos Shingles, Slate & Sheathing Company, 1906), 1; and Robert G. Skerrett, "Asbestos in Architecture," *SA*, December 4, 1920, pp. 572, 583.

104. Philip Carey Company, *Carey Flexible Cement Roofing: Specifications* (n.d.), n.p.; according to Johns Manufacturing Company, *Descriptive Price List: Asbestos* (1890), 12, double-thickness roofing felt was $2.50 per one hundred square feet. On asbestos-cement

shingles, see Asbestos Shingles, Slate & Sheathing Company, *Trade Priced List* (1906), 5; Keasbey & Mattison Company, *Offers to the Building Trade: Asbestos Shingles, Slates and Sheathings* (Ambler, Pa: Keasbey & Mattison Company, 1906); Johns-Manville Company, *The Roof Everlasting; J-M Transite Asbestos Fire-Proof Shingles* (New York: Johns-Manville Company, 1912); Mohawk Asbestos Slate Company, *Mohawk Tapered Asbestos Shingles* (Utica, N.Y.: Mohawk Asbestos Slate Company, 1923); National Asbestos Manufacturing Company, *National Asbestos and Asphalt Slate Surface Shingles; Roof Service and Fire Security at Moderate Costs* (Jersey City, N.J.: National Asbestos Manufacturing Company, 1922); and Philip Carey Company, *The Everlasting House Top: Carey Asbestos Shingles* (Cincinnati, Ohio: Philip Carey Company, 1926).

105. "Paints," *MB* 8, no. 10 (1876): 237; 9, no. 12 (1877): advertising page 1; 11, no. 10 (1879): 236; and 11, no. 11 (1879): 260; and "H. W. Johns Asbestos Liquid Paints," *American Missionary* 32, no. 5 (1878): 158.

106. Experiments were made in London with the use of asbestos paint for the fireproofing of theater sets and scenery; see "Asbestos Fire-Proof Paint," *Knowledge*, January 27, 1882, p. 271.

107. Wolfgang Venerand, *Asbest und Feuerschutz; Enthaltend: Vorkommen, Verarbeitung und Anwendung des Asbestes, Sowie den Feuerschutz in Theatern, Offentlichen Gebäuden U.S.W.; A. Hartleben's Chemisch-Technische Bibliothek; Bd. 133* (Wien, Germany: Hartleben, 1886), 108.

108. James Jackson Jarves, "American Museums of Art," *Scribner's Monthly* 18, no. 3 (1879): 407.

109. "Fire-Resisting Materials," *Science* 18, no. 458 (1891): 268.

CHAPTER 3 THE RISE OF THE ASBESTOS CURTAIN

1. Edwin O. Sachs, *Modern Opera Houses and Theatres* (London: Batsford, 1897), 3:87–119.

2. William Paul Gerhard, *Theatre Fires and Panics: Their Causes and Prevention* (New York: Wiley, 1896), 23; and Sachs, *Modern Opera Houses*, 3:128.

3. Wermiel, *Fireproof Building*, 186–218. Model codes include *Proposed Building Law for Medium Sized Cities, as Drafted by a Commission Appointed Pursuant to Chapter 579, Laws of 1892 of New York State . . . Issued June, 1893, by the Committee on Construction of Buildings of the National Board of Fire Underwriters* (New York: Committee on Construction of Buildings of the National Board of Fire Underwriters, 1893); National Board of Fire Underwriters and NFPA, *An Ordinance Providing for Fire Limits and the Construction and Equipment of Buildings in Small Towns and Villages* (New York: National Board of Fire Underwriters and NFPA, 1914); Texas State Fire Insurance Commission, *Building Code (Advisory) for Towns and Small Cities* (Austin: Texas State Fire Insurance Commission, 1922); and International Conference of Building Officials, *Uniform Building Code* (Pasadena, Calif.: International Conference of Building Officials, 1958). State codes typically covered buildings outside municipal fire limits; see, for example, New York State Board of Standards and Appeals, *Fire-Resistive Construction: Effective December 30, 1955: Industrial Code Rule No. 7* (Albany, N.Y.: New York State Board of Standards and Appeals, 1955).

4. The current editions of these titles are NFPA, *Fire Protection Handbook*, 19th ed. (2003); and NFPA, *Life Safety Code* 2000 ed. The handbook has been published since 1896; the *Life Safety Code* was incorporated into it until 1978, when the code began to be published as a separate volume.

5. This is not intended as a complete mathematical model of risk. For such a model, see Yacov Y. Haimes, *Risk Modeling, Assessment, and Management* (New York: Wiley, 1998).
6. In this chapter and elsewhere in the book, both American and British spellings of the word *theater/theatre* are used. This is because the official and historical names of many theaters, including American ones such as the Iroquois, are spelled according to the British usage; of course, authors of different nationalities and temperaments also use different spellings.
7. Donald C. Mullin, *The Development of the Playhouse; a Survey of Theatre Architecture from the Renaissance to the Present* (Berkeley: University of California Press, 1970), 82–83. See also Richard Leacroft, *Theatre and Playhouse: An Illustrated Survey of Theatre Building from Ancient Greece to the Present Day* (London: Methuen, 1984), 92; and Simon Tidworth, *Theatres: An Architectural and Cultural History* (New York: Praeger, 1973), 126.
8. Pierre Patte, *Essai sur l'architecture théatrale* (Paris: Chez Moutard, 1782), 112–13.
9. Adrien Louis Lusson, *Projet d'un théâtre d'opéra définitif pour la ville de Paris, en remplacement de l'opéra provisoire, et recherches sur le lieu propre à son érection et les causes du déplacement actuel de la population aisée de la capitale* (Paris: Gratiot, 1846), 15–16.
10. August Fölsch, *Theaterbrände und die Zur Verhütung Derselben Erforderlichen Schutz-Massregelu; Mit Einem Verzeichniss von 523 Abgebrannten Theatern* (Hamburg, Germany: Meissner, 1878), 17–50; Gerhard, *Theatre Fires and Panics*, 5; and Andrew Martin Hayes, "The Iroquois Theatre Fire: Chicago's Other Great Fire" (Ph.D. diss., University of Nebraska, 2000), 11–18.
11. F.W.J. Hemmings, "Fires and Fire Precautions in the French Theatre," *Theatre Research International* 16, no. 3 (1991): 242.
12. Frederick Howard, Earl of Carlisle, *Thoughts upon the Present Condition of the Stage, and upon the Construction of a New Theatre* (London: Clarke, 1809), 8; see also a similar view almost a century later in Horace Seal, "At the Play, before Your Own Gas-Fire," *Westminster Review* 158 (July 1902): 80; and the accounts of A. E. Wilson in *The Lyceum* (London: Yates, 1952), 29, 51–54, 195.
13. "History of Fire Safety Legislation and Other Interesting Dates," *FireNet*, *http://www.fire.org.uk/fpact/leg.htm*, accessed on May 18, 2003.
14. Wermiel, *Fireproof Building*, 11. Some American cities, such as Fredericksburg, Virginia, did not take this step until the nineteenth century—after a major fire, of course. See Edward Alvey, *The Fredericksburg Fire of 1807* (Fredericksburg, Va.: Historic Fredericksburg Foundation, 1988), 4, 13.
15. For details of these procedures, see Sol Cornberg and Emanuel Lawrence Gebauer, *A Stage Crew Handbook*, rev. ed. (New York: Harper, 1957), 169–201.
16. "Pigments through the Ages," *WebExhibits*, *http://webexhibits.org/pigments/indiv/technical/vermilion.html*, accessed on March 8, 2003. On the ordinary toxic gases associated with structural fires, see Ken Miller and Andrew Chang, "Acute Inhalation Injury," *Emergency Medicine Clinics of North America* 21, no. 2 (2003): 541–47.
17. A lustre, or chandelier, of this type is depicted in F. A. Buerki, *Stagecraft for Nonprofessionals*, 3d ed. (Madison: University of Wisconsin Press, 1972), 76.
18. Richard Allen Willis, "Killer Fires in the American Theatre," *Theater Design and Technology* 28 (February 1972): 5–6.
19. Patte, *Essai sur l'architecture théatrale,* 165–73; George Saunders, *A Treatise on Theatres* (London: Saunders, 1790; reprint, New York: Blom, 1968), 21, 30. As late as 1925, architects were still extolling the acoustical superiority of wood for theater interiors. See

Irving Pichel, *Modern Theatres* (New York: Harcourt Brace, 1925), photo caption facing pp. 24 and 32–33, in which the author praises both wooden paneling for auditorium walls and soft wood for the stage floor.

20. Bryan Akers, *Graphic Description of the Burning of the Richmond Theatre, Compiled from the Lips of Eyewitnesses* (Lynchburg, Va.: News Book and Job Office Print, 1879), 5.

21. The foregoing narrative relies on the following sources: Akers, *Graphic Description*; "Theatre on fire. AWFUL CALAMITY!" [broadside] (1812); *Collection of Facts and Statements Relative to the Fatal Event Which Occurred at the Theatre in Richmond, on the 26th December, 1811,* 2d ed. (Richmond, Va.: Dutton, 1812); *Calamity at Richmond, Being a Narrative of the Affecting Circumstances Attending the Awful Conflagration of the Theatre, in the City of Richmond, on the Night of Thursday, the 26th of December, 1811. By Which, More Than Seventy of Its Valuable Citizens Suddenly Lost Their Lives, and Many Others Were Greatly Injured and Maimed*, 2d ed. (Philadelphia: Watson, 1812); and Gynger Cook, "Richmond Theater Fire: December 26, 1811," *Shaputis, http://jshaputis.tripod.com/ClayArticles/richmond_theater_fire.htm,* accessed on February 1, 2003.

22. John Ripley Freeman, *On the Safeguarding of Life in Theaters, Being a Study from the Standpoint of an Engineer* (United States: n.p., 1906); and Sir Eyre Massey Shaw, *Fires in Theatres* (London, 1876). See also Shaw's autobiography, edited by his son Frank: *The Business of a Fireman* (London: Lomax Erskine, 1945).

23. Richard Allen Willis, "Fire at the Theatre: Theatre Conflagrations in the United States from 1798 to 1950" (Ph.D. diss., Northwestern University, 1967), 307; and *Schenk v. United States,* 249 U.S. 47, 1919.

24. Gerhard, *Theatre Fires and Panics,* 14–15. A somewhat different account appears in *A Thrilling Personal Experience! Brooklyn's Horror; Wholesale Holocaust at the Brooklyn, New York, Theatre, on the Night of December 5th, 1876; Three Hundred Men, Women and Children Buried in the Blazing Ruins! Origin, Progress and Devastation of the Fire* (Philadelphia: Barclay, 1877).

25. On the significance of this fire, see Chris Luebkeman, *A History of Structural Concrete: The Development of Structural Form, http://darkwing.uoregon.edu/~struct/resources/resources_index.html,* accessed on May 11, 2003.

26. This account of the Ring Theatre fire relies on the following sources: "Wien-Vienna Geschichte: Ringtheaterbrand, 1881," *http://www.wien-vienna.at/ringtheaterbrand.htm,* accessed on January 16, 2004; Gerhard, *Theatre Fires and Panics*, 15–16; "Modern Stage Mechanism," *SA* 81, no. 15 (1899): 232; Sachs, *Modern Opera Houses and Theatres,* 3:134; and Carl Lautenschlaeger, "Theatre Fire and Its Prevention in Germany," *SA* 90, no. 4 (1904): 59.

27. Carlisle, *Thoughts upon the Present Condition of the Stage,* 40

28. Saunders, *Treatise on Theatres*, 41.

29. "Protection against Fire in Theaters," *Forum* 36 (October 1904): 273.

30. Sachs, *Modern Opera Houses,* 3:114. For a more detailed account of these experiments, see [Fire Commissioner] Westphalen, "Experimental Theater Conflagrations," *SA Supplement* 62, no. 1615 (1906): 25873–74.

31. James George Buckley, *Theatre Construction and Maintenance: A Compendium of Useful Hints and Suggestions on the Subjects of Planning, Construction, Lighting, Fire Prevention, and the General Structural Arrangements of a Model Theatre* (London: "The Stage" Office, 1888), 29–30.

32. Charles John Hexamer, "How Theatres Should Be Constructed," *American Architect and Building News* 36 (1892): 182; and Gerhard, *Theatre Fires and Panics,* 33–35.

33. William D. Baker, "Improvement in Fire-Proof Curtains," U.S. patent no. 186657 (issued on January 30, 1877); Lemuel W. Wright, "Improvement in Fire-Shields for Theaters, &c.," U.S. patent no. 187340 (issued on February 13, 1877); Charles Hoffman, "Improvement in Theater-Curtains," U.S. patent no. 0188136 (issued on March 6, 1877); Luke Edward Jenkins, "Fire-Proof Curtain," U.S. patent no. 0397648 (issued on February 12, 1889); George W. Putnam, "Fire-Curtain," U.S. patent no. 0520218 (issued on May 22, 1894); and Sachs, *Modern Opera Houses,* 3:115.
34. Sachs, *Modern Opera Houses,* 3:117; see also U.K. Parliament, House of Commons, Select Committee on Theatres and Places of Entertainment, *Report from the Select Committee on Theatres and Places of Entertainment: Together with the Proceedings of the Committee, Minutes of Evidence, Appendix, and Index. Ordered, by the House of Commons, to Be Printed, 2 June 1892* (London: Eyre and Spottiswoode, 1892), 408, on metal-wire curtains in French theaters.
35. Gerhard, *Theatre Fires and Panics,* 75–76, 78. See also p. 107 in this work; and the same author's *Safety of Theatre Audiences and the Stage Personnel against Danger from Fire and Panic: A Paper,* Publications of the British Fire Prevention Committee no. 41 (London: British Fire Prevention Committee, 1899), 28. This type of iron curtain is illustrated in "Iron Theater Curtain," *MB* 16, no. 8 (1884): 187.
36. Buckley, *Theatre Construction and Maintenance,* 92. On the issue of sprinklers over the stage, see also Carl Lautenschlaeger, "Theatrical Engineering Past and Present," *SA Supplement* 60, no. 1542 (1905): 24703.
37. Hemmings, "Fires and Fire Precautions in the French Theatre," 244–45. Hemmings believes that the fear of fire in theaters reduced the numbers of playgoers so significantly as to have been a contributing cause of the *crise de théâtres* of the last decade of the nineteenth century (248).
38. Many European (particularly German) theaters continued to use iron curtains in the twentieth century despite these problems; see, for example, Max Littmann, *Das Stadttheater in Hildesheim* (Munich: Werner, 1909), 17; the Parisian authorities were still specifying iron for safety curtains (*rideau de fer*) in 1927, although the *décors* were supposed to be of *amiante* (asbestos). See Roger Poulain, *Salles de spectacles et d'auditions; documents recueillis et présentées* (Paris: Fréal, 1927), 7, 9, 35.
39. This account of the Exeter fire is drawn from Sachs, *Modern Opera Houses,* 3:136; and from David Anderson, *The Exeter Theatre Fire* (Royston, England: Entertainment Technology Press, 2002).
40. "Fires which Caused Theatre Disasters—a Sombre Record," *National Police Gazette* 84 (1904): 6.
41. Edwin O. Sachs, Horace S. Folker, and Ellis Marsland, *The Record of the Special Commission Formed by the British Fire Prevention Committee to Visit the Principal Cities of Central Europe, on the Occasion of the International Fire Service Congress at Budapesth, 1904* (London: British Fire Prevention Committee, 1905), 32–49; see also W. D. King, "When Theater Becomes History: Final Curtains on the Victorian Stage," *Victorian Studies* 36, no. 1 (1992): 53.
42. James Boyd, "Asbestos and Its Applications," *Journal of the Society of Arts* 34 (1886): 589.
43. "Production of Grand Opera," *SA* 76, no. 22 (1897): 1.
44. William P. Gerhard, "Can the Theater Fire be Prevented?" *SA* 90, no. 2 (1904): 22.
45. Edwin O. Sachs, *The Fire at the Iroquois Theatre, Chicago, 30th December, 1903: Comprising a Descriptive Paper by Edwin O. Sachs; the New Theatre Regulations at Chicago, the Verdict at the Inquest, with Some Constructional Particulars by the NFPA, U.S.A.,* Publications

of the British Fire Prevention Committee no. 81 (London: British Fire Prevention Committee, 1904), 9; see also 40, 44, 46.

46. This was not unusual before building codes began specifying how much asbestos was to be used in theater fire curtains. See A. Frederick Collins, "Making Asbestos Fire Curtains and Fireproof Fabrics," *SA* 90, no. 6 (1904): 122.

47. Edwin Fitzgerald Foy (1856–1928) was a popular vaudeville comedian. Bob Hope plays him in the 1955 film biography *The Seven Little Foys,* which includes the Iroquois episode. Foy threw his son Bryan, who was in the audience, to an escaping stagehand before stepping out in front of the curtain. His son Eddie Foy, Jr., and grandson Eddie Foy III also worked in theater and film.

48. "To Test Asbestos Curtains," *New York Times,* January 3, 1904, p. 2; see also Marshall Everett, *The Great Chicago Theater Disaster: The Complete Story Told by the Survivors: Presenting a Vivid Picture, Both by Pen and Camera, of One of the Greatest Fire Horrors of Modern Times* (Chicago: Publishers Union of America, 1904), 232.

49 "Means Employed in New York Theatres for Reducing Fire Hazard," *American City* 13 (October 1915): 290.

50. This account of the Iroquois Theatre fire relies on the following sources: Nat Brandt, *Chicago Death Trap: The Iroquois Theatre Fire of 1903* (Carbondale: Southern Illinois University Press, 2003); Hatch, *Tinder Box*; S. B. Hamilton, *A Short History of the Structural Fire Protection of Buildings Particularly in England,* National Building Studies special report no. 27 (London: His Majesty's Stationery Office, 1958), 32; Sachs, *The Fire at the Iroquois Theatre*; Hayes, "The Iroquois Theatre Fire"; Chicago Fire Marshall, "Iroquois Theater Fire," annual report for 1903 (1904), 10; and Dorothy Brown, "Disaster: The Iroquois Theater Fire," *Clerk of the Circuit Court, Cook County [Illinois], www.cookcountyclerkofcourt.org/Archives_/Famous_Cases/Iroquois_Theater,* accessed on January 20, 2003. On steel shutters, see "Self-Coiling Revolving Shutters," *MB* 9, no. 2 (1877): 42; and August E. Schmidt, "Safety Shutter and Screen for Shop-Windows," U.S. patent no. 0579282 (March 23, 1897); on steel and asbestos curtains in Chicago theaters, see Johns-Manville Company, *J-M Building Materials Catalog No. 102* (New York: Johns-Manville Company, 1909), 57–58; see also the company's *Johns-Manville Building Materials* (New York: Johns-Manville Company, 1920), 3.

51. Taxpayer, *To Voters: On Tuesday, June 14, 1904, You Will Have an Opportunity to Decide the Question, Whether or Not the City of Staunton Shall Close Its Opera House, or So Remodel It as to Make It as Safe for Public Use in Case of Fire or Panic, as Any Theatre in This State or Elsewhere* (Staunton, Va., 1904).

52. Freeman, *On the Safeguarding of Life in Theaters*, 41. Freeman believes the Iroquois curtain was made of asbestos, but of an inferior quality. He also endorsed the use of steel mesh or backing with the asbestos fabric (54).

53. Keasbey & Mattison Company, *Descriptive List of Keasbey & Mattison Company—Ambler Asbestos Products* (Ambler, Pa.: Keasbey & Mattison Company, 1924), 10.

54. Dayton, Ohio, Building Code Committee, *Building Code of the City of Dayton, Ordinance No. 10383* (Dayton, Ohio: Building Code Committee, 1917), 36.

55. The expression *skeuomorph* has been proposed for the linguistic survival of an obsolete technology: for example, "icebox" for "refrigerator" and "carriage return" for the "enter" key on a computer keyboard. I prefer the neologism *archaeomorph* (in Greek "old form") because *skeuomorph* has a well-established meaning in archaeology, referring to "material metaphors" rather than those in purely linguistic use. See Nicholas Gessler, "Skeuomorphs and Cultural Algorithms," *UCLA Anthropology, Computational*

Evolution, and Ecology Group, http://www.streamlive.com/, accessed on October 7, 2003. Referring to the connection tone of a telephone as a "dial tone" is an example of an archaeomorph. See also "Safety Curtain," in "Theater Terminology," *Houston Tidelanders Barbershop Chorus, http://www.swd.org./tidelanders/Scripts/TheaterGlossary.htm*, accessed on March 12, 2003. When Eddie Foy was attempting to calm the audience from the stage of the Iroquois, he wondered to himself, "Didn't anyone know how to bring that damned iron curtain down?" Hatch, *Tinder Box*, 86.

56. "Report of Committee on Safety to Life," in *Proceedings of the NFPA Thirty-third Annual Meeting* (Boston: NFPA, 1929), 348; and Noland D. Mitchell, *Tests of Theatre-Proscenium Curtains* (Washington, D.C.: U.S. Department of Commerce, Bureau of Standards, 1933).

57. There is an excellent illustration of a J. R. Clancy steel and asbestos curtain, including the driving mechanism and hydraulic check, in Harold Burris-Meyer and Edward Cyrus Cole, *Theatres & Auditoriums* (New York: Reinhold, 1949), 148. For an exceptionally handsome example of a steel curtain, see Mary L. Gray, "Lane Technical High School, Indian Motif, 1936 or 1937," in *A Guide to Chicago's Murals* (Chicago: University of Chicago Press, 2001), 488. For an example of a Sanborn map with notation of an asbestos curtain, see Sanborn-Perris Map Company, *Insurance Maps of Wilmington, Delaware* (New York: Sanborn-Perris Map Company, 1927), 1:1. These maps also note which structures have asbestos roofing.

58. Marshall K. Rouse, *Fundamentals of Fire Prevention and Protection* (Stanford, Calif.: Stanford University, Engineering, Science, and Management Defense Training Program, 1942), unpaginated section on theater fire safety; and Arthur Sherman Meloy, *Theatres and Motion Picture Houses: A Practical Treatise on the Proper Planning and Construction of Such Buildings, and Containing Useful Suggestions, Rules, and Data for the Benefit of Architects, Prospective Owners, Etc.* (New York: Architects' Supply and Publishing Company, 1916), 70. See also "The Cannonsburg Holocaust," *NFPA Quarterly* 5, no. 2 (1911): 135–36.

59. For accounts of asbestos curtains in place, see *The Players: A Private Gentlemen's Amateur Theater Club—The Last One to Leave (Lock-up), http://www.theplayers-detroit.org/index.html,* accessed on March 10, 2003; see also City of Prescott, Arizona, *Elks Opera House History Timeline, http://www.cityofprescott.net/Elks%20Opera%20House/history.cfm,* accessed on March 14, 2003, The University of Minnesota Libraries, Manuscripts Division, holds an important collection of theater curtains made by the Twin City Scenic Company between 1895 and 1979, including several decorated asbestos curtains; see, for example, the library webpage, *http://snuffy.lib.umn.edu/image/srch/bin/DisplayImage/MSSC0080,* accessed on March 10, 2003. Painted safety curtains are still used in some European and British theaters; see Kara Walker, *Safety Curtain* (Cologne, Germany: Konig, 2002).

60. For examples of modern theater safety curtains, see "Fabric Safety Curtain (Simple)," *Tatra UK Ltd., http://www.theatrestyle.com/tatra/t_safety.htm;* "Fire Resistant Cloth," *Harkness Hall USA, http://secure.room2.co.uk/harknesshall/stage.htm#7;* and "Types of Fire Safety Curtains," *J. R. Clancy, http://www.jrclancy.com/Products/firecurtaintypes.htm,* all three accessed on September 19, 2003.

61. Tidworth, *Theatres*, 184. Leacroft makes essentially the same point in *Theatre and Playhouse*, 160.

62. "Theater Fires," *http://www.disaster-management.net/theater_fire.htm,* accessed on August 15, 2004; Massimo Livi-Bacci, *A Concise History of World Population*, 3d ed. (Malden, Mass.: Blackwell, 2001), 27.

63. Commonwealth of Pennsylvania, Department of Labor and Industry, *Regulations for Protection from Fire and Panic: Class II Buildings. Theatres and Motion Picture Theatres,* revision of 1929 ed. (Harrisburg: Commonwealth of Pennsylvania, Department of Labor and Industry, 1952), 36–38.

64. C. C. Shepherd, "Proscenium-Arch," U.S. patent no. 0773228 (issued on October 25, 1904); Clarence E. Uncapher and Alden Andrus, "Fire-Curtain," U.S. patent no. 0764083 (issued on July 5, 1904); Howard M. Smith, "Fire-Shield for Proscenium-Arch Openings of Theaters," U.S. patent no. 0775865 (issued on November 22, 1904) and reviewed in "A Fire Shield," *SA* 94, no. 25 (1906), 518; John R. Clancy, "Auxiliary Release for Theatrical Fire-Curtain," U.S. patent no. 0784835 (issued on March 14, 1905); Lincoln J. Carter, "Fireproof Curtain," U.S. patent no. 0828532 (issued on August 14, 1906); Thomas B. Jackson and H. C. Yatter, "Fire-Shutter," U.S. patent no. 0829339 (issued on August 21, 1906); and Anton Karst, "Safety Appliance for Theaters," U.S. patent no. 0840692 (issued on January 8, 1907).

65. This account relies on Schneider, *Midwinter Mourning,* 7–11; and *Opera House Fire, January 13, 1908, Boyertown, Berks County, Pennsylvania: With a List of Those Who Perished in This Fire,* rev. ed. (Jamison, Pa.: Will-Britt, 1988), unpaginated, reprinted from *Genealogist's Post* 5 (February 1968).

66. Richard Ruedy, *Moving-Picture Regulations in the Provinces and Hazards of Moving-Picture Film: Provisional Report* (Ottawa: NRC, 1932), 14.

67. Thomas McIlvaine, Jr., "Reducing Film Fires," *Annals of the American Academy of Political and Social Science* 128 (November 1926): 96–99.

68. The Cinematograph Act of 1909 [U.K.], 9 Edw.7 c.30.

69. National Board of Fire Underwriters, *Rules and Requirements of the Board Governing the Storage and Handling of Nitro-Cellulose Films in Connection with Motion Picture Film Exchanges* (New York: National Board of Fire Underwriters, 1910); "Motion-Picture Films," *NFPA Quarterly* 5, no. 2 (1911): 134; and NFPA and ANSI, *Standard for the Storage and Handling of Cellulose Nitrate Motion Picture Film* (Quincy, Mass.: NFPA, 1982). This standard is updated every few years.

70. John H. Winship, "Safety from Fire in Motion Picture Theaters," *American City* 15 (December 1916): 700–701.

71. Meloy, *Theatres and Motion Picture Houses,* 59–60.

72. "A Satisfactory Fire," *NFPA Quarterly* 5, no. 1 (July 1911): 10; "Miscellaneous Fires: H-6682 Moving Picture Exhibition Hall," *NFPA Quarterly* 5, no. 2 (1911): 243; and F. H. Richardson, *Motion Picture Handbook; A Guide for Managers and Operators of Motion Picture Theatres,* 3d ed. (New York: Moving Picture World, 1916), 72, 89, 90, 211, 217, 221–22, 226, 229, 233, 244, 271. The illustration on page 244 recalls the post-fire projection booth in Giuseppe Tornatore's Italian film *Cinema Paradiso.*

73. This account is drawn from the following sources: Denis O'Shaughnessy, *Limerick: 100 Stories of the Century* (Limerick, Ireland: Leader Print, 2000), 124–28; Hugh Clevely, *Famous Fires; Notable Conflagrations on Land, Sea, and in the Air. None of Which Should Ever Have Happened* (New York: Day, 1957), 44–45; and the *Limerick Leader,* September 6–11, 1926, front section.

74. John S. Bellamy II, "Breath of Death: The 1929 Cleveland Clinic Disaster," *Timeline* 19, no. 4 (2002): 20–31; see also the role of nitrocellulose film in Washington Surveying and Rating Bureau, "Fatal Paint Factory Fire, Seattle," *NFPA Quarterly* 30, no. 3 (1937): 242–46.

75. Ruedy, *Moving-Picture Regulations*, 8.

76. For example, NFPA, "Report of Committee on Hazardous Chemicals and Explosives. Division II. Regulations for the Storage and Handling of Nitrocellulose Motion Picture Film," in *Proceedings of the NFPA Thirty-fifth Annual Meeting* (Boston: NFPA, 1931), 333–50.

77. Pennsylvania, *Regulations for Protection from Fire and Panic*, 32–35.

78. James N. Tidwell, "Adam's Off Ox: A Study in the Exactness of the Inexact," *Journal of American Folklore* 66, no. 262 (1953): 293. On the origin of the expression, see Keasbey & Mattison, *Legends of Asbestos*, 18.

79. Pacific Coast Building Officials' Conference, *Uniform Building Code* (Los Angeles: Pacific Coast Building Officials' Conference, 1937), 247–54.

80. Reinhold A. Dorwart, "Prussian Fire Protection 300 Years Ago," *NFPA Quarterly* 51, no. 3 (1958): 195–205; and James C. Robertson. *Introduction to Fire Prevention*, 4th ed. (Englewood Cliffs, N.J.: Prentice Hall, 1995), chap. 1.

81. "Individual Responsibility for Fires in France," *NFPA Quarterly* 5, no. 3 (1912): 279–81.

82. "Public Education," *NFPA Quarterly* 5, no. 3 (1912): 269; and Fire Commission of the City of Elmira, N.Y., *Building Code of the City of Elmira, N.Y., Effective July 25, 1923*, 43, 54–57, 109, 122–29.

83. J. Grove Smith, *Fire Waste in Canada* (Ottawa: Canadian Commission of Conservation, 1918), 104. For an example of timber-industry views on this point in the early twentieth century, see "Unjust to Wooden Shingles," editorial in the *Mississippi Valley Lumberman*, November 10, 1911, quoted in *NFPA Quarterly* 5, no. 3 (1912): 285. For international fire-safety professionals' views of these commodities, see National Board of Fire Underwriters, *Shingle Roofs As Conflagration Spreaders; An Appeal to the Civil Authorities and Civil and Commercial Bodies* (New York: National Board of Fire Underwriters, 1916); and the virtually identical British Fire Prevention Committee, *The Dangers of Combustible Roof Coverings: Shingle Roofs As Conflagration Spreaders, Being Some Lessons for the British Possessions Overseas*, "Red Books" of the British Fire Prevention Committee, no. 209 (London: British Fire Prevention Committee, 1917).

84. Everett Uberto Crosby and Henry Anthony Fiske, *NFPA Handbook of Fire Protection*, 8th ed. (Boston: NFPA, 1935), 379, 989; see also "Wooden Shingle Roofs: A Survey of the Present Picture in the United States," *NFPA Quarterly* 30, no. 4 (1937): 312–13.

85. Edwin O. Sachs, *A Record of the Baltimore Conflagration, February 7 and 8, 1904* (London: British Fire Prevention Committee, 1904). See also Roebling Construction Company, *The Baltimore Fire; A Description of the Effects of the Fire upon Various Materials Employed in Modern Fire-Proof Buildings; the Efficiency of the Fire-Proofing Materials; the Lessons of the Fire. The Iroquois Theatre Fire; A Description of the Building; the Story of the Fire; the Causes of the Disaster; Safeguards and Recommendations Relating to the Construction, Equipment and Management of Theatres* (New York: Roebling Construction Company, 1905).

86. Frederic R. Farrow, *"Fire-Resisting" Floors Used in London* (London: British Fire Protection Committee, 1898); "A Ceiling by the Asbestos & Asbestic Company, Ltd., London," in Edwin O. Sachs, *Facts on Fire Prevention* (London: Batsford, 1902): 1:103–9. See also the following sources, all written and published by the British Fire Prevention Committee: *Fire Tests with Partitions and Walls: A Three-Inch Partition and a Nine-Inch Wall Formed of Asbestic Bricks, and Erected by the Asbestic and Tile Company, Limited*, "Red Books" of the British Fire Prevention Committee, no. 102 (1905); *Fire Tests with Doors:*

Two Ferro-Asbestic Doors. No. 1. An Iron-Cased Composite Door Panelled and Moulded Both Sides, and No. 2. An Iron-Cased Composite Door, the External Faces Having Vertical Indentations Similar to V-Jointed Boarding, by T. Houghton & Co. (1906); *Fire Tests with Doors: An Asbestos Lined "Armoured" Door, Being a Sliding Door of Three Thicknesses of Deal, Covered Externally (Armoured) with Asbestos Board and Tinned Steel Plates*, "Red Books" of the British Fire Prevention Committee, no. 146 (1909); *Fire Tests with Doors: An Asbestos-Lined "Armoured" Door, Being a Hinged Door of Two Thicknesses of Deal Covered Externally (Armoured) with Asbestos Board and Tinned Steel Plates (without Frame); an Iron Door, Being a Hinged Door of 1/4-In. Iron Plates with Styles and Rails 4-Ins. Wide on Both Sides (with Iron Frame)*, "Red Books" of the British Fire Prevention Committee, no. 147 (1909); *Fire Tests with Doors: Two Ferro-Asbestic Doors Termed "Dreadnought Doors." No. 1, an Iron-Cased Composite Door Panelled Both Sides and No. 2, an Iron-Cased Composite Door, the External Faces Having Vertical V-Shaped Grooves on Surface; Submitted for Test by Fireproof Doors, Ltd.*, "Red Books" of the British Fire Prevention Committee, no. 149 (1910); and *Fire Tests with Roof Coverings of Asbestos Cement Corrugated Sheets, Submitted for Test by the Asbestos Manufactures Co., Ltd., London* (1913).

87. National Board of Fire Underwriters, *Building Code Recommended by the National Board of Fire Underwriters; An Ordinance Providing for Fire Limits, and Regulations Governing the Construction, Alteration, Equipment, or Removal of Buildings or Structures* (New York: Kempster, 1907), 110–15; *Building Code; A Compilation of Building Regulations Covering Every Phase of Municipal Building Activity with Special Emphasis on Fire Preventive Features* (Chicago: American School of Correspondence, 1913), 30–35; Spokane Building Code Committee, *Building Code of the City of Spokane, Washington* (Spokane, Wa.: Spokane Building Code Committee, 1936), 165–66; Port Chester, N.Y., Board of Trustees, *Building Zone Ordinance; Building Code; Plumbing Code. Village of Port Chester, N.Y.* (Port Chester, N.Y.: Board of Trustees, 1927), 93–94; and Industrial Commission of Wisconsin, *Building Code for the State of Wisconsin. Effective July 29, 1942* (Madison: Industrial Commission of Wisconsin, 1942), 49–51, 139–40.

88. William D. Brush and New York City Bureau of Buildings, *Building Code of the City of New York As Amended to May 1, 1922* (New York: Brown, 1922), 100–102; City of St. Louis, *Building Code and Electrical Code, with Amending Ordinances, City of Saint Louis, 1929* (St. Louis, Mo.: Donaldson, 1929), 101–111; and Texas State Fire Insurance Commission, *Building Code (Advisory) for Towns and Small Cities* (Austin: Texas State Fire Insurance Commission, 1922), 17–18.

89. "Fire Causes. H–6821. Flats," *NFPA Quarterly* 5, no. 2 (1911): 375.

90. "Building Materials Now Made of Rock," *Textile World* 48 (November 1914): 254.

91. "Electric Light and Power Stations," *NFPA Quarterly* 5, no. 2 (1911): 201; and NFPA, "Report of Electrical Committee," in *Proceedings of the NFPA Thirty-fifth Annual Meeting*, 82.

92. Harry Chase Brearley, *A Symbol of Safety; An Interpretative Study of a Notable Institution Organized for Service—Not Profit* (Garden City, N.Y.: Doubleday Page, 1923), 6–7. The same point was made a quarter of a century later in President's Conference on Fire Prevention, *Report of the Committee on Research* (Washington, D.C.: U.S. General Printing Office, 1947), 8.

93. William H. Merrill, Jr., *Electrical Bureau of the National Board of Fire Underwriters, Index to Laboratory Reports 1 to 1000* (Chicago: Underwriters National Electric Association, 1899); David Howell Merrill and Theodore C. Combs, *Specification Documents for Building Materials and Construction* (Los Angeles: Colling and Associates, 1935), 451–59; and

District of Columbia, Board of Commissioners, *The Building Code of the District of Columbia, Adopted by the Commissioners of the District of Columbia under and by Virtue of the Authority Conferred upon Them by the Act of Congress Approved June 14, 1879. Effective July 1, 1930* (Washington, D.C.: U.S. General Printing Office, 1930), 128, 151.

94. "Asbestos Building Lumber," *NFPA Quarterly* 2, no. 4 (1909): 437; and Edward H. Keith, "Shirt, Collar and Cuff Factories and Laundries: Processes and Hazards," *NFPA Quarterly* 1, no. 1 (1907): 5–19.

95. For examples, see ASTM, "Standard Specifications for Tolerances and Test Methods for Asbestos Yarns, A.S.T.M. Designation: D 299-29," in *ASTM Standards* (Philadelphia: ASTM, 1930), 1099–1101; "Standard Specifications for Fire Tests of Building Construction and Materials, A.S.T.M. Designation: C 19-33," in *ASTM Standards* (Philadelphia: ASTM, 1933), 254–58; "Standard Specifications and Test Methods for Asbestos Tape for Electrical Purposes, A.S.T.M. Designation: D 315-33," in *ASTM Standards* (Philadelphia: ASTM, 1942), 1163–66; (1933, 1942), 576–79; and "Standard Specifications and Test Methods for Asbestos Roving for Electrical Purposes, A.S.T.M. Designation: D 375-42," in *ASTM Standards* (Philadelphia: ASTM, 1942), 568–72.

96. C.J.H. Woodbury, *The Fire Protection of Mills* (New York: Wiley, 1882), 116; and "Roofings," *NFPA Quarterly* 2, no. 3 (1909): 349.

97. H. C. Jones, Jr., "Fire Hazards of Mill Villages," in *Proceedings of the NFPA Thirty-third Annual Meeting*, 131.

98. ASTM, *ASTM E84 Standard Test Method for Surface Burning Characteristics of Building Materials* (Philadelphia: ASTM, 1958 and later editions). This standard is also known as NFPA 255 and UL 723.

99. N. B. Pope, "The Automobile As a Fire Hazard," *NFPA Quarterly* 5, no. 1 (1911): 41ff. This article recommended that the "controller" be asbestos-lagged for fire protection. On aviation, see, for example, "Aviation Fire Hazards," *NFPA Quarterly* 22, no. 2 (1928): 161–75.

100. Armand Bindi and Daniel Lefeuvre, *Le Métro de Paris: Histoire d'hier à demain* (Rennes: Ouest-France, 1990); and Johns-Manville Company, *Reducing the Weight of Car Flooring: Johns-Manville Service to Railroads* (New York: Johns-Manville Company, 1929).

101. "Babies and Matches," *NFPA Quarterly* 5, no. 3 (1912): 261–62.

102. Brearly, *Symbol of Safety*, 11.

103. Cincinnati, Ohio, City Council, *Building Code of the City of Cincinnati, Effective April 1, 1933*, vol. 51, *Code of Ordinances* (Cincinnati, Ohio: Cincinnati City Council, 1933), 93, 142–43, 206, 246, 281–82, 284, 313. There are similar provisions in Detroit, Department of Buildings and Safety Engineering, *Official Building Code of Detroit, Michigan* (Detroit: Department of Buildings and Safety Engineering, 1941); Boston, *Building Code of the City of Boston* (Boston: Printing Department, 1944); and St. Louis, *Building Code and Electrical Code*. See also Pacific Coast Building Officials Conference, *Uniform Building Code*; National Board of Fire Underwriters, *Building Code*, 5th ed.; and NRC, *National Building Code of Canada* (Ottawa: NRC, 1941).

104. Everett Uberto Crosby, *NFPA Handbook of Fire Protection*, 8th ed. (Boston: NFPA, 1936).

105. See, for examples, "Unsatisfactory or Serious Fires in Buildings Equipped with Sprinklers," *NFPA Quarterly* 5, no. 1 (1911): 78–97; and H. E. Hilton, "Why Unsatisfactory Sprinkler Fires," *NFPA Quarterly* 42, no. 3 (1949): 205–28.

106. Brearly, *Symbol of Safety*, 50.

107. Kimberly D. Rohr, *U. S. Experience with Sprinklers* (Quincy, Mass.: NFPA, 2003), 45.

108. Gene Eric Salecker, *Disaster on the Mississippi: The Sultana Explosion, April 27, 1865* (Annapolis, Md.: Naval Institute Press, 1996); Peter Maust, "'Congress Could Do Nothing Better': Promoting Steamboat Safety in Antebellum America," *Quarterly of the National Archives and Records Administration* 32, no. 2 (2000): 101–13; John K. Ward, "The Future of an Explosion," *American Heritage of Invention and Technology* 5, no. 1 (1989): 58–63; and J. G. Burke, "Bursting Boilers and the Federal Power," *Technology and Culture* 7, no. 1 (1966): 1–23.

109. The spontaneous ignition of coal is the catastrophe depicted in Joseph Conrad, "Youth: A Narrative," in *Typhoon and Other Tales* (New York: New American Library, 1962), 219–48. For other cargoes representing a similar risk, see Kenneth C. Barnaby, "Dangerous Cargoes," in *Some Ship Disasters and Their Causes* (London: Hutchinson, 1968), 227–449.

110. Samuel D. McComb, "Report of Marine Committee," in *Proceedings of the NFPA Thirty-second Annual Meeting* (Boston: NFPA, 1928), 187.

111. Edward Frank Spanner, *The Fire-Proof Ship* (Liverpool, England: Birchall, 1934), 59–60.

112. Alfred Cecil Hardy and Edward Tyrrell, *Shipbuilding: Background to a Great Industry* (London: Pitman, 1964), 66–72.

113. John Griffith Armstrong, *The Halifax Explosion and the Royal Canadian Navy: Inquiry and Intrigue* (Vancouver: UBC Press, 2002); Michael J. Bird, *The Town That Died: The True Story of the Greatest Man-Made Explosion before Hiroshima* (New York: Putnam, 1963); Bill Minutaglio, *City on Fire* (New York: HarperCollins, 2003); and Hugh W. Stephens, *The Texas City Disaster, 1947* (Austin: University of Texas Press, 1997).

114. "Report of Committee on Piers and Wharves," NFPA, in *Proceedings of the NFPA Thirty-fifth Annual Meeting*, 361; see also the association's "Report of Committee on Laws and Ordinances: Ordinance on Piers and Wharves," in *Proceedings of the NFPA Forty-fourth Annual Meeting* (Boston, NFPA, 1940), 475–82.

115. On ports, see Seattle Building Department, *Building Code of the City of Seattle, 1924* (Seattle, Wa.: Seattle Building Department, 1924), 110–11.

116. O'Donnell, *Ship Ablaze*.

117. British Fire Prevention Committee, *Constructional Fire Stops on Passenger Ships*, "Red Books" of the British Fire Prevention Committee, no. 205 (London: British Fire Prevention Committee, 1912); U.K. Board of Trade, *International Convention for the Safety of Life at Sea, 1929* (London: His Majesty's Stationery Office, 1929); *What Is "SOLAS"? http://www.uscg.mil/hq/g-m/mse4/solas.htm,* accessed on April 30, 2003; Donald V. Reardon, "Safety of Life from Fire at Sea," *NFPA Quarterly* 43, no. 1 (1949): 47–52; and R. G. Furnivall, *Fire and Fire Protection with Particular Reference to Ships* (Richmond, England: Association of Engineering and Shipbuilding Draughtsmen, 1953), 23–27.

118. Spanner, *Fire-Proof Ship,* 99–100.

119. There is a popular account of *L'Atlantique* fire in Bernard Bernadac and Claude Molteni de Villermont, *L'Incendie de l'Atlantique* (Nantes: Marines, 1997), 77–127. See also H. Gerrish Smith's summary of the effect of these fires on the international maritime community in his "Fire Protection on Sea-Going Vessels," in *Proceedings of the NFPA Fiftieth Annual Meeting* (Boston: NFPA, 1946), 124–29. On the importance of asbestos to marine electrical wiring, see Albert King and Victor Harold Wentworth, *Raw Materials for Electric Cables* (London: Benn, 1954), 322–25.

120. "Report of Marine Committee," in *Proceedings of the NFPA Thirty-fifth Annual Meeting,* 265–68; and "Report of Marine Committee," in *Proceedings of the NFPA Thirty-fourth Annual Meeting* (Boston: NFPA, 1930), 75–89.

121. Gallagher, *Fire at Sea.*

122. See, for example, Julius F. Wolff, Jr., "Ship Fires on Lake Superior," *Inland Seas* 54, nos. 1–4 (1998), especially 4:13.

123. U.K. Board of Trade, *International Convention for the Safety of Life at Sea, 1929*, at 119 pages; and U.S. Office of the President, *Safety of Life at Sea. Convention, with Regulations, between the United States of America and Other Governments, Signed at London June 10, 1948* (Washington, D.C.: U.S. Government Printing Office, 1953), at 241 pages. See also U.S. Congress, Senate Committee on Foreign Relations, *International Convention for the Safety of Life at Sea: Hearing* (Washington, D.C.: U.S. Government Printing Office, 1949), 2–9. The 1960 edition was even longer at 388 pages; for asbestos, see *International Convention for the Safety of Life at Sea* (London: Her Majesty's Stationery Office, 1965), 144.

124. William D. Winter, "Marine Insurance in a World at War," *Journal of the American Association of University Teachers of Insurance* 9, no. 1 (1942): 41, 43.

CHAPTER 4 MASS DESTRUCTION BY FIRE

1. The porcine incendiaries were directed at Moslem soldiers, no doubt in part to show contempt and render them ritually unclean, in the siege of Belgrade in 1456. See Stephen Turnbull, *The Ottoman Empire, 1326–1699* (Oxford: Osprey, 2003), 75.

2. Washington, D.C., was a village when it was burned by the British in August 1814. On Columbia, the standard secondary source is Marion Brunson Lucas, *Sherman and the Burning of Columbia* (Columbia: University of South Carolina Press, 2000); for the "horse's mouth" version, see William T. Sherman and O. O. Howard, *Who Burnt Columbia? Official Depositions of Wm. Tecumseh Sherman and Gen. O. O. Howard, U.S.A., for the Defence, and Extracts from Some of the Depositions for the Claimants* (Charleston, S.C.: Evans and Cogswell, 1873). Some Columbians might argue that I have identified the wrong end of the horse. See also Leo Tolstoy, *War and Peace,* trans. Louise and Aylmer Maude (New York: Norton, 1996), 798.

3. Harold Everett Wessman and William Allen Rose, *Aerial Bombardment Protection* (New York: Wiley, 1942), 45.

4. World War I is thought by some to be a counterexample to this principle, but see, for example, David Mitrany, *The Effect of the War in Southeastern Europe* (New Haven, Conn.: Yale University Press, 1936), 73. For a more detailed discussion of this issue, see Rachel Maines, "Wartime Allocation of Textile and Apparel Resources: Emergency Policy in the Twentieth Century," *Public Historian*, 7 no. 1 (1985): 28–51.

5. Richard S. Foster and Francis P. Hoeber, "Limited Mobilization: A Strategy for Preparedness and Deterrence in the Eighties," *Orbis* 24 (fall 1980): 451.

6. "New Worlds to Conquer," *SA* 121, no. 11 (1919): 254, 268. Demand had been rising faster than supply for more than a decade at that time, as E. Schaaf-Regelman noted in "Asbestos," *Engineering Magazine* 34 (October 1907): 68–80.

7. Stephen Leacock, *Nonsense Novels*, 3d ed. (London: Lane, 1912), 218. On the Elgin marbles, see "Proceedings of the Society for the Promotion of Hellenic Studies, Session 1918–1919," *Journal of Hellenic Studies* 39 (1919): xvii–xxxi.

8. "Contraband of War," *American Journal of International Law* 10, no. 4, supp. (1916): 19, 26, 52; "Proclamation Making Certain Further Exports Unlawful in Time of War," *American Journal of International Law* 12, no. 4, supp. (October 1918): 305; "Gradual Ascendency of the United States As an Asbestos Producer," *SA* 115, no. 23 (1916): 155;

and "Uses of Asbestos and Its Products in Chemical and Metallurgical Industries," *Metallurgical and Chemical Engineering* 12, no. 5 (1914): 356.

9. Imperial Institute, *The Mineral Industry of the British Empire and Foreign Countries. War Period. Asbestos. (1913–1919)* (London: His Majesty's Stationery Office, 1921).

10. Oliver Bowles, *Asbestos,* bulletin no. 403 (Washington, D.C.: U.S. Department of the Interior, U.S. Bureau of Mines, 1937), 8–9; "Friction Materials: Linings and Blocks for Brakes and Clutches," *Product Engineering* 12 (January 1941): 2–6; and "Automobiles and the Asbestos Industry," *Automotive Industry,* March 2, 1922, p. 520.

11. Oliver Bowles, "Asbestos," *Power Plant Engineering,* April 15, 1920, p. 418; S. R. Winters, "Asbestos in America," *SA* 123, no. 14 (1920): 325; John Morris Combs, "Roofs Everlasting of Fibrous Rock," *Arts and Decoration* 26 (January 1927): 64–66; Belle Caldwell, "Asbestos in Construction and Decoration," *Industrial Education Magazine* 30 (January 1929): 240–43; "Improved Asbestos-Cement Wallboard," *Pencil Points* 22, supp. (1941): 84; G.T.K. Norton, "Asbestos Wall Board," *House and Garden* 61 (February 1932): 70.

12. This was also discovered to be true of other roofing types, but it was a new idea in 1921. Johns-Manville Company, *Rock or Rags* (New York: Johns-Manville Company, 1920); Johns-Manville Company, *Re-Roofing for the Last Time* (New York: Johns-Manville Company, 1921); Johns-Manville Company, *Book of Roofs* (New York: Johns-Manville Company, 1923); Asbestos Shingles, Slate & Sheathing Company, *Ambler Asbestos Shingles; Types of Roofs Illustrating the Best Permanent Fireproof Roofing* (Ambler, Pa.: Asbestos Shingles, Slate & Sheathing Company, 1925). On do-it-yourself, see R. M. Bolen, "Modernize Your Bathroom," *Popular Science* 125 (September 1934): 6–7; and "Unit Wall Construction: Asbestos-Concrete," *Architectural Forum* 57, supp. (September 1932): 26.

13. On the role of asbestos pads in laundry and ironing, see Florence Buchanan, *Home Crafts of Today and Yesterday* (New York: Harper, 1917), 3; Lydia Ray Balderston's *Laundering* (Philadelphia: Balderston, 1914), 166, 195, 204; and *Laundering, Home-Institution* (Philadelphia: Lippincott, 1923), 243, 248, 292–93. On the cleaning of fur, see William E. Austin, *Principles and Practice of Fur Dressing and Fur Dyeing* (New York: Van Nostrand, 1922), 73, 80. Stove mats are endorsed in Emily Burbank, *Be Your Own Decorator* (New York: Dodd Mead, 1922), 145. Home fire insulation of hot-air pipes and radiators is discussed in Maria Parloa, *Home Economics; A Guide to Household Management Including the Proper Treatment of the Materials Entering into the Construction and the Furnishing of the House* (New York: Century, 1898), 23; and Mary Harrod Northend, *The Art of Home Decoration, Home Economics Archive—Research, Tradition and History* (New York: Dodd Mead, 1921), 51–54. Jane Fales dismisses the material as insignificant to dressmaking in *Dressmaking, a Manual for Schools and Colleges* (New York: Scribner's, 1917), 116; but Margaret Brooke and Winifred Brooke thought asbestos lace important enough to include in their *Lace in the Making: With Bobbins and Needle* (London: Routledge, 1923), 143. On dyeing, see Christopher Rawson, Walter Myers Gardner, and W. F. Laycock, *A Dictionary of Dyes, Mordants, and Other Compounds Used in Dyeing and Calico Printing* (London: Griffin, 1918), 46, 188–89, 191–96; and Thomas Franz Hanausek, Andrew Lincoln Winton, and Kate Grace Barber Winton, *The Microscopy of Technical Products* (New York: Wiley, 1907), 22, 51, 156–57. Asbestos is ideal as a dye filter partly because it cannot itself be bleached or dyed and absorbs no color from the dyestuff; see G. S. Fraps, *Principles of Dyeing* (New York: Macmillan, 1903), 2. On balloons, see Johns-Manville Company, *Johns-Manville Service to Industry: Asbestos Roofings, Heat Insulations, Packings, Electrical Insulations, Waterproofing, Industrial Flooring, Etc.* (New York: Johns-Manville Company, 1924), 141.

14. U.K. Home Office, *Fire Precautions in Schools* (London: His Majesty's Stationery Office, 1935), 23; and J. S. Trevor, "Silencing London's Tubes: Continuously Welded Rail and Wall Screens of Perforated Asbestos Wood," *Transit Journal* 82 (December 1938): 482–83.
15. Oliver Bowles and K. G. Warner, "Asbestos," in *Minerals Yearbook*, ed. H. Herbert Hughes (Washington, D.C.: Superintendent of Documents, 1939), 1309.
16. E. Willard Miller, "Some Aspects of the United States Mineral Self-Sufficiency," *Economic Geography* 23, no. 2 (1947): 82; and Myron W. Watkins, "Scarce Raw Materials: An Analysis and a Proposal," *American Economic Review* 34, no. 2 (1944): 242.
17. G. W. Taussig, "The Tariff Debate of 1909 and the New Tariff Act," *Quarterly Journal of Economics*, 24, no. 1 (1909): 36.
18. G. W. Josephson and F. M. Barsigian, "Asbestos," in *Minerals Yearbook*, ed. A. F. Matthews (Washington D.C.: Superintendent of Documents, 1949), 139–48.
19. Raymond Dumett, "Africa's Strategic Minerals During the Second World War," *Journal of African History* 26, no. 4 (1985): 399, 405.
20. J. H. Pratt, "Asbestos," in *Mineral Resources of the United States*, ed. David T. Day (Washington, D.C.: U.S. Department of the Interior, U.S. Bureau of Mines, 1903), 1114; Oliver Bowles, "Asbestos," in *Mineral Resources of the United States*, G. F. Loughlin (Washington, D.C.: U.S. Department of the Interior, U.S. Bureau of Mines, 1921), 46; Oliver Bowles, *Asbestos* (Washington, D.C.: U.S. Department of the Interior, U.S. Bureau of Mines, 1937), 7; and Oliver Bowles and F. M. Barsigian, "Asbestos," in *Minerals Yearbook*, ed. C. E. Needham (Washington, D.C.: Superintendent of Documents, 1942), 1433. See also Alexander Tarn and Robert W. Campbell, "A Comparison of U.S. and Soviet Industrial Output," *American Economic Review* 52, no. 4 (1962): 707, 12, 26.
21. Boris Baievsky, "Siberia—the Storehouse of the Future," *Economic Geography* 3, no. 2 (1927): 183.
22. Britain had been exploring for asbestos deposits not only in its own colonies but within its general sphere of influence since the late nineteenth century. See, for example, Thomas P. Brockway, "Britain and the Persian Bubble, 1888–92," *Journal of Modern History* 13, no. 1 (1941): 42.
23. Asbestos and Mineral Corporation, *Asbestos from Mine to Finished Product* (New York: Asbestos and Mineral Corporation, 1919), 123.
24. Oliver Bowles and B. H. Stoddard, "Asbestos," in *Minerals Yearbook*, ed. O. E. Kiessling (Washington, D.C.: Superintendent of Documents, 1935), 1117; and (1936), ed. Kiessling, 995.
25. Robert E. Cryor, *The Asbestos Textile Industry in Germany*, Fiat final report, no. 460 (London: His Majesty's Stationery Office, 1945); see also U.S. Bureau of Mines and National Security Resources Board, Materials Office, *Materials Survey, Asbestos, 1950* (Washington, D.C.: U.S. Bureau of Mines and National Security Resources Board, 1952), part 13, sec. 4.
26. Bowles and Barsigian, "Asbestos" (1942), 1428.
27. U.S. Bureau of Mines and National Security Resources Board, *Materials Survey . . . 1950*, part 8, sec. 1; part 10, sec. 1.
28. U. S. National Recovery Administration, *Code of Fair Competition for the Asbestos Industry As Approved on November 1, 1933 by President Roosevelt* (Washington, D.C.: U.S. General Printing Office, 1933), 2:273–87.
29. J. Hurstfield, "The Control of British Raw Material Supplies, 1919–1939," *Economic History Review* 14, no. 1 (1944): 19.

30. B. K. [*sic*], "Soviet Developing Huge Asbestos Reserves," *Far Eastern Survey* 4, no. 21 (1935): 169–70; on U.S. advisers to Soviet asbestos manufacture, see Wladimir Naleszkiewicz, "Technical Assistance of American Enterprises to the Growth of the Soviet Union, 1929–1933," *Russian Review* 25, no. 1 (1966): 76; and Lewis S. Feuer, "American Travelers to the Soviet Union 1917–32: The Formation of a Component of New Deal Ideology," *American Quarterly* 14, no. 2, part 1 (1962): 138.

31. On asbestos deposits in Japan, see W. Gowland and Professor Milne [*sic*], "Exploration in the Japanese Alps, 1891–1894: Discussion," *Geographical Journal* 7, no. 2 (1896): 147. On the exploitation of Manchurian minerals in the 1930s, see John R. Stewart, "Manchoukuo Now in Third Year of Five-Year Plan," *Far Eastern Survey* 8, no. 10 (1939): 118; and Andrew J. Grajdanzev, "Manchuria: An Industrial Survey," *Pacific Affairs* 18, no. 4 (1945): 334.

32. Bowles, "Asbestos" (1936), 995.

33. Jesse M. Weaver, *Asbestos Textiles and Textile Products* (Manheim, Pa.: Raybestos-Manhattan Company, Industrial Sales Division, 1940); "Gast's Asbestos Air-Cell Covering," *SA* 79, no. 23 (1898): 357; S. H. Dodd, "Asbestos Suits That Aid in Fighting Oil Fires," *American City* 37 (August 1927): 229–30; "Report of Committee on Manufacturing Hazards: Paint Spraying and Spray Booths," in *Proceedings of the NFPA Forty-fourth Annual Meeting* (Boston: NFPA, 1940), 455–63; "Value of Sheet Asbestos on Hot Pipes," *Mechanical Engineering* 42 (January–March 1920): 69 (January), 188–89 (March); Oliver Bowles, "Asbestos," in *The Mineral Industry*, ed. G. A. Roush (New York: McGraw-Hill, 1927), 71; "Active Demand for Asbestos Wire," *Electrical World*, September 11, 1920, pp. 548–49; R. E. Forbester, "Asbestos Cement and Other Asbestos Building Products," *Chemistry and Industry*, September 2, 1939, pp. 808–12; and A. V. Staniforth, "Prevention of Corrosion in Steel Chimneys," *Engineering*, September 12, 1941, pp. 204–5. The use of asbestos as fertilizer is reported again four decades later in J. M. West and Victoria R. Schreck, "Asbestos," in *Minerals Yearbook*, ed. Charles W. Merrill (Washington, D.C.: U.S. General Printing Office, 1962), 272.

34. See the following catalogs from Johns-Manville Company: *Johns-Manville Service to Railroads; Asbestos Roofings, Packings, and Insulations: Steam and Electrical Railroad Supplies* (New York: Johns-Manville Company, 1923); *Johns-Manville Service to the Power Plant* (New York: Francis, 1931); and *Johns-Manville Service to the Oil Industry: Manuals of J-M Service to the Industry No. 103* (New York: Johns-Manville Company, 1931).

35. Emily Thompson, "Dead Rooms and Live Wires: Harvard, Hollywood, and the Deconstruction of Architectural Acoustics, 1900–1930," *Isis* 88, no. 4 (1997): 609. In some types of sound insulation service after World War II, asbestos materials were used in conjunction with fiberglass; see, for example, H. R. Hogendobler, "Fibrous Glass in the Power Plant," *Power Plant Engineering* 50 (July 1946): 89.

36. See the following catalogs from the Asbestos Buildings Company, all published by the company in Ambler, Pa.: *Fireproof Ambler Asbestos School Buildings* (1926); *Ambler Fireproof Asbestos School Buildings "Save Children's Lives"* (1925); and *Fireproof Ambler Asbestos Bungalows* (1926); also see Johns-Manville Products Corporation, *Ideas for Station Modernization: Johns-Manville Building Materials* (New York: Johns-Manville Products Corporation, 1940).

37. On Britain, see M. J. Elsas, "War and Housing," *Economic Journal* 50, no. 197 (1940): 43; and "Billets for Britain: Asbestos Hutments," *Architectural Forum* 82 (May 1945): 7. For use in the United States, see Charles O. Jackson and Charles W. Johnson, "The Urbane Frontier: The Army and the Community of Oak Ridge, Tennessee, 1942–1947," *Military*

Affairs 41, no. 1 (1977): 8, 10. On substitutions for metal, see "U.S. Rubber Fabric Saves Plane Metal; Asbestos Product Used in Heating and Ventilating Systems," *Aviation News*, April 24, 1944, p. 14; and "Asbestos Ducts Save Metal," *American Builder* 64 (April 1942): 114. On contemporary standards for the use of asbestos in ventilation systems, see Carl W. Wheelock, "Report of Committee on Air Conditioning," in *Proceedings of the NFPA Forty-fourth Annual Meeting*, 265–76.

38. Asbestos Protected Metal Company, *Asbestos Protected Metal for Roofing and Siding, Bulletin No. 55* (Pittsburgh: Asbestos Protected Metal Company, 1915), 8, 16, 22, 36; American Rolling Mill Company, *Armco Asbestos-Bonded Steel* (Middletown, Ohio: American Rolling Mill Company, 1946); L. Sonneborn Sons, *Stormtight: An Asbestic Water-Proof Compound for Covering Old and New Roofs* (New York: L. Sonneborn Sons, 1921); Preservative Products Company, *Liquid Asbestos Fibre Roof Cement* (Cleveland, Ohio: Preservative Products Company, 1921); Peerless Products Manufacturing Company, *Peerless Liquid; Equivalent to Ten Coats of Ordinary Paint; An Asbestos Liquid Cement for Renewing Your Old Roof* (New York: Peerless Products Manufacturing Company, 1922); A-Re-Co Refining Company, *"A-Re-Co" Liquid Cement; An Asbestos Product for Leaky Roofs; Contains No Coal Tar* (New York: A-Re-Co Refining Company, 1922); Continental Asbestos & Refining Company, *Stonoleum, for New Floors . . . For Old and Worn Floors . . . For Patching* (St. Louis: Continental Asbestos & Refining Company, n.d.); H. W. Johns–Manville Company, *Johns-Manville Asbestos Roofing: Corrugated Form* (New York: H. W. Johns–Manville Company, 1920), 14–15; and Johns-Manville Corporation, *Poultry Houses: Protected with Johns-Manville Building Materials: Durable, Sanitary* (New York: Johns-Manville Corporation, 1938).

39. "Stopping the Leaks," *SA* 127, no. 1 (1922): 38.

40. "New Type of Insulation," *Science Abstracts* 7 (1904): 959–60.

41. Kenneth Leith and Donald Macy Liddell, *The Mineral Reserves of the United States and Its Capacity for Production* (Washington, D.C.: National Resources Committee, 1936), 24–30.

42. Robert A. Clifton, "Asbestos," in *Minerals Yearbook* (Washington, D.C.: U.S. General Printing Office, 1970), 203; Brian Davis, "Asbestos and the Alternatives," *Engineer*, March 24, 1983, pp. 26–27; and G. L. Moses, "Fiber Glass Compared with Asbestos," *Electrical World*, April 8, 1939, p. 989.

43. On high-temperature insulation on Navy ships and the Navy's testing program, see R. R. Gurley, "Pipe Covering Materials for High Temperatures," *Journal of the American Society of Naval Engineers* 47 (May 1935): 247; D. L. Cox, "Heat Insulation," *Transactions of the Society of Naval Architects and Marine Engineers* 44 (1937): 470–86; "Asbestos Industry Reaches High Point," *Chemical and Engineering News*, April 10, 1942, 472; Graham Lee Moses, "Fiber Glass Compared with Asbestos," *Electrical World*, April 8, 1939, p. 989; and D. Wolochow and W. H. White, "Heat Resistance of Asbestos: Abstract," *Chemical and Metallurgical Engineering* 48 (May 1941): 184.

44. U. S. Tariff Commission, *Asbestos: A Summary of Information on Uses, Production, Trade and Supply*, Industrial Materials Series, no. M-3 (Washington, D.C.: U.S. Tariff Commission, 1951), 3–8; Oliver Bowles and A. C. Petron, "Asbestos," in *Minerals Yearbook*, ed. F. M. Shore (Washington, D.C.: Superintendent of Documents, 1941), 1427; see also U.S. Bureau of Mines and National Security Resources Board, *Materials Survey . . . 1950*, sec. 1; part 9, sec. 5; part 12, secs. 4–6.

45. Oliver Bowles and K. G. Warner, "Asbestos," in *Minerals Yearbook* (Washington, D.C.: Superintendent of Documents, 1940), 1363–72; U. S. Congress, Senate Committee on

Military Affairs, *Strategic and Critical Materials Hearings before the United States Senate Committee on Military Affairs, Seventy-seventh Congress, First Session, May 15, 19, 21, 26, June 4, 11, 16, July 1, 1941* (Washington, D.C.: U.S. General Printing Office, 1941); on the Harbord List, see Jules Backman, *War and Defense Economics, Containing Text of Defense Production Act Incorporating 1951 Amendments* (New York: Rinehart, 1952), 132.

46. Oliver Bowles and F. M. Barsigian, "Asbestos," in *Minerals Yearbook*, ed. C. E. Needham (Washington, D.C.: Superintendent of Documents, 1943), 1474.

47. U.S. Congress, *Strategic and Critical Materials Hearings* (1941), 7.

48. Harry N. Holmes, *Strategic Materials and National Strength* (New York,: Macmillan, 1942), 9, 73; Bowles and Barsigian, "Asbestos" (1942), 1427–28; see also Johns-Manville Company, *Johns-Manville Service to the Iron and Steel Industry, Manuals of J-M Service to Industry—No. 101* (New York: Johns-Manville Company, 1930).

49. Fortunately for U.S. asbestos supplies, it was not necessary to cover the entire length of either pipeline with this type of insulation. Bowles and Barsigian, "Asbestos" (1943), 1478.

50. U.S. Bureau of Mines and National Security Resources Board, *Materials Survey . . . 1950*, part 18, sec. 2; C. R. Hutchcroft, "Durability of Asbestos-Cement Siding," *American Concrete Institute Journal* 16 (September 1944): 116–17; on hangars, see B. M. Doolin, "Fighting Airplane Crash Fires," in *Proceedings of the NFPA Forty-seventh Annual Meeting* (Boston: NFPA, 1943), 137.

51. C. P. Balsam, "Emergency Substitutes for Metals," *Sheet Metal Worker* 32 (October 1941): 30–33; "Asbestos Ducts Save Metal," *American Builder* 64 (April 1942): 114; "U.S. Rubber Fabric Saves Plane Metal; Asbestos Product Used in Heating and Ventilating Systems," *Aviation News*, April 24, 1944, p. 14; Curtis R. Wellborn, "Substitutes," *NFPA Quarterly* 35, no. 3 (1942): 236; and Johns-Manville Corporation, *Transite Pipe, Transite Ventilators, Transite Leaders and Gutters, BMT 400* (New York: Johns-Manville Corporation, 1932–33).

52. "Priority Control on South African Asbestos," *Foreign Commerce Weekly*, February 7, 1942, 20.

53. U.S. War Production Board, "Asbestos Textiles," *Conservation Order M-123 As Amended December 14, 1942* (Washington, D.C.: U.S. War Production Board, 1942), list A.

54. Dick Hutchinson, "Cool Birdhouses of Asbestos Cement," *Popular Mechanics* 75 (1941): 590–91; and Belle Caldwell, "Asbestos in Construction and Decoration," *Industrial Education Magazine* 30 (1929): 240–43.

55. S. H. Ingberg, "Incendiary Bomb Structural Protection," *NFPA Quarterly* 35, no. 3 (1942): 232.

56. Ernest Joseph King, *U.S. Navy at War, 1941–1945: Official Reports to the Secretary of the Navy* (Washington, D.C.: U.S. Department of the Navy, 1946), 146–49.

57. For a detailed analysis of these issues, see Per Walmerdahl, *An Introduction to the Concept of Weapon-Induced Fires* (Lund, Sweden: Lund University, Department of Fire-Safety Engineering, 1999).

58. On Navy ship fires generally in World War II, see Samuel Eliot Morison, *History of United States Naval Operations in World War II* (Boston: Little Brown, 1947), 1:327–29, 7:xxix, 8:280–82, 14:95–99.

59. James S. Russell, "Design for Combat," in *Carrier Warfare in the Pacific: An Oral History Collection*, ed. E. T. Wooldridge, Smithsonian History of Aviation Series (Washington, D.C.: Smithsonian Institution Press, 1993), 91–99; see also data on *Essex*-class carriers

of the World War II era at "National Museum of Aviation," *http://naval.aviation.museum/hep/hep_carrier.html*, accessed on November 20, 2003.

60. John S. Thach, "The Big Blue Blanket," in *Carrier Warfare in the Pacific*, 265.

61. C. A. Pinkstaff, D. L. Sturtz, and R. F. Bellamy, "USS *Franklin* and the USS *Stark*—Recurrent Problems in the Prevention and Treatment of Naval Battle Casualties," *Military Medicine* 154, no. 5 (1989): 232. Many of the bodies from this action were never recovered.

62. "Lost American Aircraft Carriers," *http://members.tripod.com/~ffhiker/index-7.html*, accessed on November 22, 2003; Stephen Jurika, Jr., 'The *Franklin*: Tragedy and Triumph," in *Carrier Warfare in the Pacific*, 247–64; and Roger Chesneau, *Aircraft Carriers of the World, 1914 to the Present: An Illustrated Encyclopedia* (Annapolis, Md.: Naval Institute Press, 1984), 227. On the fire vulnerability of aircraft carriers generally, see Norman Friedman, *U.S. Aircraft Carriers: An Illustrated Design History* (Annapolis, Md.: Naval Institute Press, 1983), 139, 208, 211, 226–36, 261, 291.

63. Scot MacDonald, *Evolution of Aircraft Carriers* (Washington, D.C.: U.S. Department of the Navy, Office of the Chief of Naval Operations, 1964), 55. On the *Franklin*, see Gareth Pawlowski, *Flat-Tops and Fledglings; A History of American Aircraft Carriers* (South Brunswick, N.J.: Barnes, 1971), 119–26.

64. Roger Williams, "Safety at Sea, with Special Reference to the New S.S. *America*," in *Proceedings of the NFPA Forty-fourth Annual Meeting*, 39–43; and George S. Bull, "Wartime Activities of NFPA Marine Section," in *Proceedings of the NFPA Forty-seventh Annual Meeting*, 74–83.

65. Albert Edward Tompkins, *Marine Engineering (a Text-Book)*, 5th ed. (London: Macmillan, 1921), 106–7, 125, 135, 141–42, 147, 152, 320, 331, 511–12, 525, 792, 865; Alan Osbourne, *Modern Marine Engineer's Manual* (Cambridge, Md.: Cornell Maritime Press, 1945), chap. 4, pp. 37 76–83, 107; chap. 5, p. 11; chap. 7, p. 146; chap. 8, pp. 49, 82–85; Alan Osbourne and A. Bayne Neild, *Modern Marine Engineer's Manual*, 2d ed., vol. 1 (Cambridge, Md.: Cornell Maritime Press, 1965), chap. 3, pp. 49–67, chap. 19, pp. 3–15; see also Frank Rushbrook, *Fire Aboard; The Problems of Prevention and Control in Ships and Port Installations* (New York: Simmons-Boardman, 1961), 257, 332–47.

66. I am, of course, by no means the first to suggest asbestos as suitable apparel for visiting the infernal regions. See, for example, Lawrence Daniel Fogg, *The Asbestos Society of Sinners, Detailing the Diversions of Dives and Others on the Playground of Pluto, with Some Broken Threads of Drop-Stitch History, Picked up by a Newspaper Man in Hades and Woven into a Stygian Nights' Entertainment* (Boston: Mayhew, 1906); and Gordon Challis (b. 1932), "The Asbestos-Suited Man in Hell," in *Anthology of Twentieth Century New Zealand Poetry*, ed. Vincent O'Sullivan (London: Oxford University Press, 1970), 295–96. George Washington makes a cameo appearance in Fogg's novel, dressed, like the other characters, in an asbestos suit.

67. A different type of suit, with separate asbestos trousers, hood, gloves, and boots, can be seen in S. H. Dodd, "Asbestos Suits That Aid in Fighting Oil Fires," *American City* 37 (August 1927): 229–30. Oilfield firefighter Ford Alexander is reported to have worn a suit of this kind. The cumbersome qualities of early proximity suits and their headgear are evident in the illustration in Ralph Howard, "Role of Asbestos," *SA* 123 (1920): 42.

68. Eugene W. Nelson, "Cloth of Stone," *Nature Magazine* 39 (December 1946): 528; see also "Asbestos: The Magic Protection," *Safety Engineering* 96 (July 1948): 11–13, 48–49.

69. The old style of suit is illustrated in Johns-Manville Company, *Johns-Manville Service to the Iron and Steel Industry* (1930), 53; for the new models, see "Asbestos Suit Designed for Rapid Use," *SA* 167, no. 2 (1942): 73. The tests are described in "Modern Fire-walkers," *Popular Mechanics* 66 (November 1936): 641–42, 152A, 154A. For the British term, see Alfred J. S. Bennett, *Ship Fire Prevention* (Oxford: Pergamon, 1964), 60. On the introduction of zippers, see Robert D. Friedel, *Zipper: An Exploration in Novelty* (New York: Norton, 1994).
70. "U.S. Air Force Studies on Protective Clothing," *NFPA Quarterly* 42, no. 4 (1949): 265–68; Charles R. Schmitt, *Pyrophoric Materials Handbook: Flammable Metals and Materials* (1996), *http://www.towson.edu/~schmitt/pyro/chapter3.html,* accessed on May 3, 2004. Schmitt observes of burning aluminum powder that "the intense heat generated in this flame makes it impossible for a person to stand close to the flame for more than a few seconds, and ordinarily the [cutting] torch operator must wear an asbestos suit."
71. L. W. Bauer, "Package for Electrolyte; Leakproof Container for Sulfuric Acid," *Modern Packaging* 19 (September 1945), 129, 170, 172.
72. The exemption of inland waterways from regulations applying to seagoing ships was a factor in the most fatal of all U.S. single-vessel maritime disasters: the explosion and fire on the riverboat *Sultana* in April 1865, in which more than 1,700 lives were lost. Gene Eric Salecker, *Disaster on the Mississippi: The Sultana Explosion, April 27, 1865* (Annapolis, Md.: Naval Institute Press, 1996), 59, 206.
73. "The Noronic Pyre," *NFPA Quarterly* 43, no. 3 (1950): 213–26.
74. These two accounts are drawn from Wynn F. Foster, *Fire on the Hangar Deck: Ordeal of the Oriskany* (Annapolis, Md.: Naval Institute Press, 2001); and Gregory A. Freeman, *Sailors to the End: The Deadly Fire on the U.S.S. Forrestal and the Heroes Who Fought It* (New York: Morrow, 2002).
75. See, for examples, Margaret R. Hayes and Alice M. Stoll, *Fleet Evaluation of Fire Resistant Flight Clothing, Phase I* (Warminster, Pa.: Naval Air Development Center, 1963); B. T. Lee, *Naval Shipboard Fire Risk Criteria—Berthing Compartment Fire Study and Fire Performance Guidelines*, ed. W. J. Parker (Washington, D.C.: U.S. Department of the Navy, Center for Fire Research, 1976), 44; B. T. Lee, *Naval Performance Testing of Bulkhead Insulation Systems for High Strength to Weight Ship Structures: Final Report, August 1976* (Washington, D.C.: U.S. Department of Commerce, National Bureau of Standards, 1976); B. T. Lee, *Fire Performance Guidelines for Shipboard Interior Finish*, ed. W. J. Parker (Washington, D.C.: U.S. Department of the Navy, Center for Fire Research, 1979), 16, 18; B. T. Lee and W. J. Parker, *Fire Buildup in Shipboard Compartments: Characterization of Some Vulnerable Spaces and the Status of Prediction Analysis: Final Report*, NBSIR 79-1714 (Washington, D.C.: U.S. Department of Commerce, National Bureau of Standards, 1979), 5–6, 30–40; and B. T. Lee, *Fire Hazard Evaluation of Shipboard Hull Insulation and Documentation of a Quarter-Scale Room Fire Test Protocol*, NBSIR 83-2642 (Washington, D.C.: U.S. Department of Commerce, National Bureau of Standards, 1983).
76. Tessa Richards, "Medical Lessons from the Falklands," *British Medical Journal* 286, no. 6367 (1983): 790–92; David Brown, *The Royal Navy and the Falklands War* (London: Cooper, 1987), 140–50, 194–214; Max Arthur, *Above All, Courage: The Falklands Front Line: First-Hand Accounts* (London: Sidgwick and Jackson, 1985), 29–42, 95–135; and Paul Eddy, Magnus Linklater, and Peter Gillman, *The Falklands War* (London: Deutsch, 1982), 156–67.
77. Pinkstaff et al., "U.S.S. *Franklin* and the U.S.S. *Stark*," 229–33.

78. Asbestos and Magnesia Manufacturing Company, *Ehrets 85% Magnesia Coverings* (1905); see also Kidde advertisement, "Motorbike Built for Two," in *Aviation*, 43 no. 5 (1944): 105.

79. "Asbestos in Air Raid Precautions," *Electrician*, July 7, 1939, p. 23; Horatio Bond, *Fire Defense; A Compilation of Available Material on Air-Set Fires, Bombs and Sabotage, Civilian Defense, Fire Fighting and the Safeguarding of Industrial Production for Defense* (Boston: NFPA, 1941), 15; and John West, Jr., "Total Fire—the Incendiary Bomb," *NFPA Quarterly* 34, no. 2 (1940): 102–24. On the British civilian defense pamphlet, see James M. Landis, "Fire Protection in Civilian Defense," in *Proceedings of the NFPA Forty-sixth Annual Meeting* (Boston: NFPA, 1942), 56.

80. James K. McElroy, "The Work of the Fire Protection Engineers in Planning Fire Attacks," in *Fire and the Air War*, 133.

81. Most of this account is from U.S. Strategic Bombing Survey [hereafter USSBS], *Area Studies Division Report: European War No. 31* (Washington, D.C.: USSBS, Civilian Defense Division, January 1947), 3, 9, 21, app. A; and USSBS, *Over-all Report (European War)* (Washington, D.C.: USSBS, Civilian Defense Division, September 30, 1945), 3–4 in the *Summary* volume (Washington, D.C.: USSBS, Civilian Defense Division, 1947), 72, 74; 91–101 in the longer version; see also R. J. Overy, *Why the Allies Won* (New York: Norton, 1995), 109–33.

82. USSBS, Civilian Defense Division, *Final Report: European War, No. 40* (Washington, D.C.: USSBS, Civilian Defense Division, January 1947), 51, 150–51.

83. R. R. Waesche, "Port Fire Safety," in *Proceedings of the NFPA Forty-sixth Annual Meeting*, 60–65. Waesche was a commandant in the Coast Guard. See also the remarks of NFPA General Manager Percy Bugbee and those of A. J. Smith, "Prevention and Control of Marine Fires," in the same volume, 49–52 and 190–93 respectively.

84. USSBS, Civilian Defense Division, *Hamburg Field Report: European War No. 44* (Washington, D.C.: USSBS, Civilian Defense Division, January 1947), 1:32–33; see also 1:38, 41, 70.

85. U.S. Air University, Air Force Historical Division, Research Studies Division, *Historical Analysis of the 14–15 February 1945 Bombings of Dresden, http://www.airforce-history.hq.af.mil/PopTopics/dresden.htm,* accessed on November 22, 2003.

86. Jerome Bernard Cohen, *Japan's Economy in War and Reconstruction* (Minneapolis: University of Minnesota Press, 1949), chap. 1.

87. Robert Guillain, *I Saw Tokyo Burning: An Eyewitness Narrative from Pearl Harbor to Hiroshima* (Garden City, N.Y.: Doubleday, 1981), 176–77. Some of this gray tile must have been the asbestos roofing and sheathing that was widely used on industrial buildings in Japan.

88. Caidin, *A Torch to the Enemy*, 185; USSBS, Physical Damage Division, *Effects of Incendiary Bomb Attacks on Japan: A Report on Eight Cities* (Washington, D.C.: USSBS, 1947), 87–97; USSBS, Physical Damage Division, *A Report on Physical Damage in Japan* (Washington, D.C.: USSBS, 1947), 95; and Kerr, *Flames Over Tokyo*, 139–214.

89. Karl T. Compton, "Address of Dr. Karl T. Compton," in *Proceedings of the NFPA Fiftieth Annual Meeting* (Boston: NFPA, 1946), 34.

90. USBBS, *Report on Physical Damage in Japan*, 82, 166–67. For photographs and other observations on asbestos use in wartime Japan, see 43, 89, 114, 165, 175, 189, 202; and USSBS, *Effects of Incendiary Bomb Attacks on Japan*, 1, 6, 10, 14, 20–23, 31, 67, 73, 80–81, 87, 102, 115, 129, 163, 194.

91. USSBS, *Report on Physical Damage in Japan*, 207. Wessman and Rose did not think asbestos made good shielding against splinters; see their *Aerial Bombardment Protection*, 349–50.

92. David J. Price and Hylton R. Brown, "Report of Committee on Dust Hazards," in *Proceedings of the NFPA Forty-sixth Annual Meeting*, 139; and Carl von Clausewitz, *On War*, trans. Michael Howard and Peter Paret (Princeton, N.J.: Princeton University Press, 1976), 119.

93. John Griffith Armstrong, *The Halifax Explosion and the Royal Canadian Navy: Inquiry and Intrigue, Studies in Canadian Military History* (Vancouver: University of British Columbia Press, 2002), 37, 42, 54, 63.

94. "Hoboken Pier Fire," *NFPA Quarterly* 23, no. 4 (1930): 37–40; and "Hoboken Pier Fire," *NFPA Quarterly* 38, no. 2 (1944): 112–18.

95. "The Normandie Fire," *NFPA Quarterly* 36, no. 1 (1942): 52–68; and *The President's Conference on Fire Prevention* [program] (Washington, D.C.: U.S. General Printing Office, 1947), 19; see also "Salvage of the French Liner: Normandie," *http://bigfishmiami.com/pages/salvage.htm*, accessed on December 3, 2003.

96. On the draw down of local fire personnel, see Peter Steinkellner, "Problems of Fire Department Operation in Wartime," in *Proceedings of the NFPA Forty-seventh Annual Meeting*, 123–27; on U.S. fire professionals in Britain see Horatio Bond's "Fire and War" in the same volume, 51–58; on underwriters, see Lamar C. Sledge, "Fire Prevention at War Department Installations," in *Proceedings of the NFPA Forty-eighth Annual Meeting* (Boston: NFPA, 1944), 65–68.

97. John Montague Thomas, *A Better America, Not a New One* (New York: National Board of Fire Underwriters, 1944), 13.

98. Eliot Ness, "Cleveland Fire Defense Plan," in *Proceedings of the NFPA Forty-fifth Annual Meeting* (Boston: NFPA, 1941), 196.

99. Stephen G. Badger and Robert S. McCarthy define a catastrophic multiple-death fire as one that kills "five or more people in a residential property, or three or more in a nonresidential or nonstructural property." See their "Catastrophic Multiple-Death Fires," *NFPA Journal* 97, no. 5 (2003): 70–75.

100. Unlike some cities, Natchez had had a code since at least 1905; see G.B.C. Shields, *Code of Ordinances of the City of Natchez* (Natchez, Miss.: Natchez Print. and Stationery Company, 1905).

101. NFPA, *Fire Investigations: Dance Hall Fire (Rhythm Club), Natchez, Mi, April 23, 1940*, [fire investigation report] (Quincy, Mass.: NFPA, 1976); and "Natchez Dance Hall Holocaust," *NFPA Quarterly* 34, no. 1 (1940): 70–75.

102. This account of the Cocoanut Grove fire is drawn from the following sources: Benzaquin, *Fire in Boston's Cocoanut Grove*; Edward Keyes, *Cocoanut Grove* (New York: Atheneum, 1984); Robert Selden Moulton, Casey C. Grant, Doug Beller, and Jennifer Sapochetti, *The Cocoanut Grove Night Club Fire, Boston, November 28, 1942* (Boston: NFPA, 1943); "Catastrophe: Boston's Worst," *Time*, December 7, 1942, p. 26; "Catastrophe," *Newsweek*, December 7, 1942, pp. 43–44; "Boston Fire: Record Death Toll Reaches 492 As 119 Survivors Struggle for Life," *Life* 13, no. 24 (1942): 47–48; Bernard DeVoto, "The Easy Chair: After Cocoanut Grove," *Harper's Magazine* 186 (1943): 335–36; Robert S. Moulton, "Remarks [on Cocoanut Grove Fire]," in *Proceedings of the NFPA Forty-seventh Annual Meeting*, 24–27; and articles in the *Boston Globe*, November 29–-December 4, 1943.

103. Moulton et al., *Cocoanut Grove Night Club Fire*, 7. See also Doug Beller and Jennifer Sapochetti, "Searching for Answers to the Cocoanut Grove Fire of 1942," *NFPA Journal* 94, no. 3 (2000): 84–92. For a discussion of Pearl Harbor as a fire emergency, see W. W. Blaisdell, "Honolulu Air Raid Fire," *NFPA Quarterly* 35, no. 3 (1942): 221–25.

104. These treatments made combustible fabrics such as cotton more difficult to ignite, but how effective they could have been on cotton soaked in gasoline and paraffin was open to question even at the time. See, for example, Warren Y. Kimball, "Hartford Circus Holocaust," *NFPA Quarterly* 38, no. 1 (1944): 9–20.

105. This account relies on the following sources: Henry S. Cohn and David Bollier, *The Great Hartford Circus Fire: Creative Settlement of Mass Disasters* (New Haven, Conn.: Yale University Press, 1991); Stewart O'Nan, *The Circus Fire: A True Story* (New York: Doubleday, 2000); "Hartford History: The Circus Fire," *http://www.hartfordhistory.net/circusfire.html,* accessed on November 29, 2003.

106. Sherwood Brockwell, "Practical Aspects of Law Enforcement for Life Safety," in *Proceedings of the NFPA Fifty-first Annual Meeting* (Boston: NFPA, 1947, 69–73.

107. "Terminal Hotel Fire, Atlanta, GA," *NFPA Quarterly* 32, no. 1 (1938): 46–51; "Hotel Plaza Fire, Jersey City," *NFPA Quarterly* 31, no. 4 (1938): 314–19; "Queen Hotel Fire, Halifax, Nova Scotia," *NFPA Quarterly* 32, no. 4 (1939): 310–18; and "Marlborough Hotel Fire, Minneapolis," *NFPA Quarterly* 33, no. 4 (1940): 384–87.

108. "St. John's Hostel Fire," *NFPA Quarterly* 36, no. 3 (1943): 187–90; and Janet McNaughton, "Knights of Columbus Hostel Fire," *http://www.avalon.nf.ca/~janetmcn/fire.html,* accessed on November 29, 2003.

109. H. Oram Smith, "Gulf Hotel Fire," *NFPA Quarterly* 37, no. 2 (1943): 83–92; and "Richmond Hotel Fire," *NFPA Quarterly* 37, no. 4 (1944): 315–17.

110. James K. McElroy, "The Lasalle Hotel Fire," and "Dubuque, Iowa, Hotel Fire," *NFPA Quarterly* 40, no. 1 (1946): 4–18 and 19–25 respectively.

111. The account that follows relies primarily on Heys and Goodwin, *The Winecoff Fire;* and James K. McElroy, "The Hotel Winecoff Fire," *NFPA Quarterly* 40, no. 3 (1947): 140–59.

112. *NFPA Handbook of Fire Protection*, 10th ed. (Boston: NFPA, 1948), 561–62.

113. *NFPA Fire Protection Handbook*, 17th ed. (Quincy, Mass.: NFPA, 1991), 6–39. ASTM established the tentative standard in 1950 and NFPA in 1953. See also A. J. Steiner, "Fire Hazard Tests of Building Materials," *NFPA Quarterly* 37, no. 1 (1943): 69–78.

114. *The President's Conference on Fire Prevention*, 2, 4.

115. *Final Report of the Continuing Committee of the President's Conference on Fire Prevention*, House document, 80th Congress, report no. 19720-6 (Washington, D.C.: U.S. General Printing Office, 1948), 12.

116. *Report of the Committee on Building Construction, Operation, and Protection of the President's Conference on Fire Prevention* (Washington, D.C.: U.S. General Printing Office, 1947), 2, 4, 9.

CHAPTER 5 SCHOOLS, HOMES, AND WORKPLACES

1. "Krause Milling Co. Explosion and Fire," *NFPA Quarterly* 31, no. 1 (1937): 59–66; and David J. Price and Hylton R. Brown, "Report of Committee on Dust Explosion Hazards," in *Proceedings of the NFPA Forty-sixth Annual Meeting*, 138–39.

2. See, for example, Jack A. Bradley, "Rehabilitation of Transportation in Western Germany," *Economic Geography* 25, no. 3 (1949): 180–89.

3. *A Fire-Resistance Test on a Structural Steel Column Protected with 5/16 In. Sprayed Limpet Asbestos* (Boreham Wood: U.K. Department of Scientific and Industrial Research and Fire Offices' Committee, Joint Fire Research Organisation, 1953); *A Fire-Resistance Test on a Structural Steel Column Protected with Sprayed "Limpet" Asbestos Grade L. B. 9* (Boreham Wood: U.K. Department of Scientific and Industrial Research and Fire Offices' Committee, Joint Fire Research Organisation, 1957); U.K. Department of the Environment and Joint Fire Research Organization, *Fire-Resistance of Floors and Ceilings* (London: Her Majesty's Stationery Office, 1961); U.K. Fire Service Department, *Building Construction,* vol. 8, *Manual of Firemanship; A Survey of the Science of Firefighting*, 2d ed. (London: Her Majesty's Stationery Office, 1963); R. Baldwin and P. H. Thomas, *The Spread of Fire in Buildings—The Effect of the Type of Construction,* Fire Research Note, no. 735 (Boreham Wood, U.K.: Fire Research Station, 1968); and J. A. Clarke, "Some Features in the Planning and Construction of Buildings from a Fire-Risk Standpoint," *Institution of Fire Engineers Collected Papers* (New York: Institution of Fire Engineers, 1945), vol. 1, part 5, pp. 18–19.
4. *New Zealand Standard Specification for Mineral Fibre Cement, Unreinforced Flat Sheets and Corrugated Sheets Superseding Edition of November, 1939* (September 1950); *New Zealand Standard Specification: By-Law for Precautions against Fire and Panic in Buildings for Public Meetings Etc.* (February 1951); *New Zealand Standard Specification for Thermal Insulating Materials for Buildings and Report on the Application of Thermal Insulation to Buildings* (June 1959); and *New Zealand Standard Model Building Bylaw* (December 1963), 51–52; all published in Wellington by the New Zealand Standards Institute, Department of Industries and Commerce.
5. Gary Snyder, *Stockpiling Strategic Materials* (San Francisco: Chandler, 1964), 140–49.
6. NFPA, *Building Exits Code* (Boston: NFPA, 1927, 1966); see also Paul E. Teague, "Case Histories: Fires Influencing the *Life Safety Code*," supp. 1 of *Life Safety Code Handbook* (Quincy, Mass.: NFPA, 2000).
7. The account that follows relies on Von Drehle, *Triangle*; Stein, *Triangle Fire;* and the Cornell University, Kheel Center, "Triangle Fire," *http://www.ilr.cornell.edu/trianglefire/default.html,* accessed on December 6, 2003.
8. Asbestos panels were routinely used in explosion testing by fire-safety professionals. See, for example, James J. Duggan, "Venting of Tanks Exposed to Fire," *NFPA Quarterly* 37, no. 2 (1943): 132–53.
9. J. W. Roberts Ltd. Midland Works, *Spray: A New and Simple Process for the Application by Spray of Limpet Asbestos* (Leeds, U.K.: J. W. Roberts, 1935); and Keasbey & Mattison Company, *K & M, Best in Asbestos* (Ambler, Pa.: Keasbey & Mattison Company, 1933–45).
10. D. M. Baird, "Montreal High-Rise Fire," *NFPA Quarterly* 57, no. 2 (1963): 119–25.
11. "Sprayed-Asbestos Technique Cuts Reinsulation Time by 55 Per Cent," *Electrical World,* August 27, 1962, p. 66; "Sprayed Asbestos Insulation Permits Coating to Be Applied Joint-Free," *Oil and Gas Journal,* May 23, 1966, pp. 145–46; British Standards Institution, *Sprayed Asbestos Insulation, British Standard 3590, 1963* (London: British Standards Institution, 1963); and Melvin S. Abrams, "Behavior of Inorganic Materials in Fire," in *Design of Buildings for Fire Safety*, ed. T. Z. Harmathy (Ottawa: NRC, 1978), 70–71.
12. See, for example, the remarks of Dorothy Mees, "Fire Prevention Is for Women, Too," *Proceedings of the NFPA Fifty-first Annual Meeting*, 53–55.
13. "1000 School Fires," *NFPA Quarterly* 33, no. 2 (1939): 145–214.

14. This account of the Lakeview Elementary School fire is drawn primarily from "The Collinwood School Fire," in *The Encyclopedia of Cleveland History* (1998), *http://www.ech.cwru.edu/ech-cgi/article.pl?id=CSF*, accessed on December 5, 2003; Marshall Everett [Henry Neil], *Complete Story of the Collinwood School Disaster and How Such Horrors Can Be Prevented, Memorable Fires of the Past* (Cleveland, Ohio: Hamilton, 1908). There were articles about the fire in March 1908 in the *Chicago News*, the *Cleveland Leader*, the *Hartford Courant*, and the *New York World*.
15. John Wesley Sahlstrom, *Some Code Controls of School Building Construction in American Cities; An Evaluation of Certain Building Code Requirements* (New York: Columbia University, Teachers College, 1933), 66–67.
16. Crosby, *NFPA Handbook of Fire Protection* (1935), 310.
17. H. W. Forster, *Fire Protection for Schools* (Boston: U.S. Department of the Interior, Bureau of Education, 1919), 16–17, 34, 38; and Merle A. Stoneman, Knute Oscar Broady, and Alanson D. Brainard, *Planning and Modernizing the School Plant* (Lincoln: University of Nebraska Press, 1949), 191, 277.
18. Clara Elizabeth Hinson Woodson, "Cleveland School Tragedy, May 17, 1923," *Update: The Newsletter of the Kershaw County Historical Society*, May 17, 1998, pp. 1–9.
19. For examples of the normative literature of school fire safety before 1960, see NFPA, *Safeguarding School Children from Fire* (Boston: NFPA, 1916); George Drayton Strayer and Nickolaus Louis Engelhardt, *Standards for Elementary School Buildings* (New York: Columbia University, Teachers College, 1933), 27, 29, 49, 57; S. F. Tirrell, "School Buildings Need Protection." *NFPA Quarterly* 38, no. 3 (1945): 216–18; *Fire Safety in the Planning of School Buildings* (Albany: State University of New York, State Education Department, Division of School Buildings and Grounds, 1947); Nelson E. Viles, *School Buildings: Remodeling, Rehabilitation, Modernization, Repair* (Washington, D.C.: Federal Security Agency, Office of Education, 1950); and his *School Fire Safety*, bulletin 1951, no. 13 (Washington, D.C.: Federal Security Agency, Office of Education, 1951); Helen K. MacKintosh, "Are Today's Children Safe from Fire?" *School Life* 33, no. 5 (1951): 74–77; Frank G. Lopez, "School Design in 1952," *Architectural Record* 111 (January 1952): 138–41; Frank H. Gage, "Fire Safety in Older School Buildings," *American School Board Journal* 129, no. 6 (1954): 42–43; National Board of Fire Underwriters, "Inspect Your School for Fire Safety," *School Life* 34, no. 9 (1952): 138–39; Samuel Miller Brownell, "School Buildings Should Be Safe," *School Life* 36 (May 1954): inside front cover.
20. "Nineteen Large Loss School Fires," *NFPA Quarterly* 51, no. 4 (1958): 275–78.
21. Richard E. Stevens, "Flames Trap Fifteen in One-Story School," *NFPA Quarterly* 47, no. 4 (1954): 331–37.
22. Alanson D. Brainard, *Handbook for School Custodians*, 4th ed. (Lincoln: University of Nebraska Press, 1952), 140–203.
23. The account that follows relies mainly on Cowan and Kuenster, *To Sleep with the Angels*; and Chester I. Babcock and Rexford Wilson, "The Chicago School Fire," *NFPA Quarterly* 52, no. 3 (1959): 155–75. A class B door provides one and a half to two hours of fire resistance and was adequate for most stairway doors that did not protect openings in fire-division walls.
24. John M. Wise, "Retroactive Application of Fire Laws," *NFPA Quarterly* 55, no. 2 (1961): 111–15; and Vincent M. Brannigan, "Record of Appellate Courts on Retrospective Fire-safety Codes," *Fire Journal* 75, no. 6 (1981): 62–72.
25. National Council on Schoolhouse Construction, *NCSC Guide for Planning School Plants*

(East Lansing, Mich.: National Council on Schoolhouse Construction, 1964), 37, 67, 71, 95.

26. Robert S. Moulton, "Combustible Fibreboards," *NFPA Quarterly* 43, no. 2 (1949): 83–95; and "Combustible Fibreboard Tile Ceiling Fire," *NFPA Quarterly* 50, no. 2 (1956): 127–29. This second reference concerns a fire in the Moncton, New Brunswick, City Hospital in May 1956.

27. Peter J. Hearst, *Wall Materials for Hospital Corridors* (Port Hueneme, Calif.: Naval Construction Battalion Center, Civil Engineering Laboratory, Bureau of Medicine and Surgery, 1980), 9–10.

28. James R. Bell, "Fourteen Die in Ohio Boarding Home Fire," *Fire Journal* 74, no. 4 (1980): 28–31, 87; and "Fires in Two Boarding Facilities Kill 34 Residents," *Fire Journal* 76, no. 4 (1982): 44–57.

29. T. Seddon Duke, "Automatic Sprinklers for Life Safety in Institutions," *NFPA Quarterly* 57, no. 3 (1964): 267–71.

30. "Hospital and Institution Fires," *NFPA Quarterly* 22, no. 3 (1929): 296–340.

31. William E. Patterson, "Little Sisters of the Poor Home Fire, Pittsburgh," *NFPA Quarterly* 25, no. 2 (1931): 111–18.

32. "Auburn, Maine, Baby Home Fire," *NFPA Quarterly* 38, no. 4 (1945): 288; and "Hartford Christmas Tree Tragedy," *NFPA Quarterly* 39, no. 3 (1946): 242–43.

33. James K. McElroy, "The Tragedy of St. Anthony Hospital," *NFPA Quarterly* 43, no. 1 (1949): 12–33.

34. "Davenport, Iowa, Hospital Fire," *NFPA Quarterly* 43, no. 3 (1950): 144–47; "Twenty Patients Dead in Nursing Home Fire," *NFPA Quarterly* 46, no. 3 (1953): 172–75; Chester I. Babcock, "Florida Nursing Home Fire," *NFPA Quarterly* 46, no. 4 (1953): 293–300; Ernest Juillerat, Jr., "The Hartford Hospital Fire," *NFPA Quarterly* 55, no. 3 (1962): 295–303; and Sidney Wohlfeld, "But It Was Fireproof!," *Safety Maintenance* 135 (August 1966): 32–35. For later accounts of fires of this type, see Robert F. Hamm, "Maples Convalescent Home Fire," *Fire Journal* 59, no. 2 (1965): 5–9; Albert B. Sears, Jr., "Nursing Home Fire, Marietta, Ohio," *Fire Journal* 64, no. 3 (1970): 5–9; and David P. Demers, "25 Die in Nursing Home," *Fire Journal* 75, no. 1 (1981): 30–35. On nursing home fires in general, see Chester I. Babcock, "A Place for Old Folks to Live," *NFPA Quarterly* 51, no.1 (1957): 35–49. Noncombustible finishes, including the mineral ceiling tile, functioned exactly as designed in a fire at the Howard University Hospital in 1977; see Richard Best, "Howard University Hospital Fire, Washington D.C.: Successful Construction and Reactions Prevent Fatalities," *Fire Journal* 75, no. 1 (1981): 30–35.

35. "Official Responsibility," *NFPA Quarterly* 7, no. 2 (1913): 134–35.

36. James R. Bell, "Twenty-Nine Die in Biloxi, Mississippi Jail Fire," *Fire Journal* 77, no. 6 (1983): 44–55.

37. Richard Best, *Reconstruction of a Tragedy: The Beverly Hills Supper Club Fire, Southgate, Kentucky, May 28, 1977* (Washington, D.C.: U.S. Department of Commerce, National Fire Prevention and Control Administration and National Bureau of Standards, 1986); Richard Best and David P. Demers, *Investigation Report on the MGM Grand Hotel Fire, Las Vegas, Nevada, November 21, 1980* (Quincy, Mass.: NFPA, 1982); and *Lodging Executive's Guide to Modern Hotel Fire Safety* (Port Washington, N.Y.: Rusting, 1985), 6.

38. "What Is Asbestos Good For?" *Mill and Factory* 47 (November 1950): 99–102.

39. U. S. Bureau of Labor Statistics, *Occupational Outlook Handbook* (Washington, D.C.: U.S. General Printing Office, 1951), 387–88; (1957), 278–80. By the 1957 edition, wages had

increased to a range between $3.15 per hour in Alabama and $3.85 per hour in New York City.

40. These mats were used in school chemical laboratories as well, although presumably without the crocheted decoration. See Regina Burrichter, Sr., "On Burning Holes in Teacher's Desk," *History Teacher* 4, no. 3 (1971): 57–58. On later concerns about the health hazards of using asbestos in dentistry, see J. R. Legan et al., "Biologic Exposure to Dental Materials," *Oral Surgery, Oral Medicine, and Oral Pathology* 36 no. 6 (1973): 908–14.

41. "Matching Sunflower Set," *Workbasket* 15, no. 12 (1950): 29–31; Spool Cotton Company, "Quick-as-a-Wink Gifts: Hot Plate Mat Cover," in *Ideas for Gifts*, no. 225 (New York: Spool Cotton Company, 1949), 2; "Pansy Hot Dish Mat," *Workbasket* 16, no. 8 (1951): 4–5; "Dishtowels from Asbestos," *Science News Letter*, September 25, 1948, p. 204; and "CPSC and Milton Bradley Co. Recall 'Fibro-Clay,'" in *News from CPSC, U.S. Consumer Product Safety Commission*, release no. 83-012. (Washington, D.C.: U.S. Consumer Product Safety Commission, March 1983).

42. J. G. Ross and G. F. Jenkins, "Asbestos," in *Industrial Minerals and Rocks (Nonmetallics Other Than Fuels)*, ed. Samuel Hood Dolbear, Harry Foster Bain, and Walter Wadsworth Bradley (New York: American Institute of Mining and Metallurgical Engineers, Committee on the Industrial Minerals, 1937), 93.

43. Alan B. Mountjoy, "The Development of Industry in Egypt," *Economic Geography* 28, no. 3 (1952): 225; P. C. Merritt, "California Asbestos Goes to Market," *Mining Engineering* 14 (September 1962): 57–60.

44. Robert B. Fisher, R. L. Thorne, and H. C. Van Cott, "Paligorskite, A Possible Asbestos Substitute," *U.S. Bureau of Mines Informational Circular* 7313 (1945): 5; "Mountain Leather May Have Many Commercial Uses," *Science News Letter*, April 28, 1945, p. 265; and G. D. Hough, "Tremolite Asbestos Now Mined in Arctic," *Mining and Metallurgy* 26 (July 1945): 364.

45. Keasbey & Mattison Company, *K & M, Best in Asbestos*; and Asbestos Wood Manufacturing Corporation, *Eterno Products; Asbestos Wood, Asbestos Lumber, Insulating Board, Ebony Board; Catalog and Price List* (Poughkeepsie, N.Y.: Asbestos Wood Manufacturing Corporation, 1923), 3, 14–15.

46. Timothy C. May, "Asbestos," in *Minerals Yearbook* (Washington, D.C.: U.S. General Printing Office, 1965), 209–10; and Dominick V. Rosato, "Asbestos-Phenolics Aid Rocket and Missile Flight," *Plastics World* 16 (April 1958): 4–6.

47. A. B. Cummins, "Fillers: Asbestos and Diatomaceous Silica," *Modern Plastics* 17 (October 1939): 51–52; H. J. Tapsell, "Second Report of the Pipe Flanges Research Committee: Investigation of Asbestos Packing," *Institution of Mechanical Engineers Journal and Proceedings* 141 (1939): 433–38; H. E. Barkan, "Superior Heat Resistance and Bonding of Asbestos-Base Phenolic Compounds," *Electrical Manufacturer* 61 (May 1958): 12; Carroll E. Shaw, "Supervision of Teen-Age Rocketeers," *NFPA Quarterly*, 52, no. 2 (1958): 99–103; Donald O. Kennedy and James M. Foley, "Asbestos," in *Minerals Yearbook*, ed. Charles W. Merrill (Washington, D.C.: U.S. General Printing Office, 1959), 200; Timothy C. May, "Asbestos," in *Minerals Yearbook* (Washington, D.C.: U.S. General Printing Office, 1963), 262; "Five Step Fabricating Method Makes a Tough Flame Shield," *Materials in Design Engineering* 61 (April 1965): 21, 23; "Fire-Safe Ball Valve Result of Asbestos-TFE," *Materials in Design Engineering* 59 (March 1964): 172; "Asbestos-Rubber Gasket Material Has Excellent Sealing Properties," *Materials in Design Engineering* 64 (December 1966): 23, 25; R. G. Walmsley, "Asbestos-Based

Bearing Materials," *British Plastics* 36 (July 1963): 407–8; and "Fire Curtains in Ten Enginehouses: Norfolk & Western Installs Barriers of Asbestos-Cement Sheets," *Railway Age*, January 27, 1945, pp. 225–26.

48. "Asbestosis among Workers in the Asbestos Textile Industry," *Public Health Reports* 53 (1938): 1893–94; "New Jersey Passes Silicosis and Asbestosis Law," *Rock Products* 48 (January 1945): 228–30; Anthony J. Lanza, William J. McConnell, and J. William Fehnel, "Effects of the Inhalation of Asbestos Dust on the Lungs of Asbestos Workers," *Public Health Reports* 50, no. 1 (1935): 1–12; American Conference of Governmental Industrial Hygienists, "Threshold Limit Values for 1950," *Archives of Industrial Hygiene and Occupational Medicine* 2 (April 1950): 98–100; and Jerry Mitchell, "Health Progress in Asbestos Textile Works," *Archives of Environmental Health* 3 (July 1961): 43–47. On dust control outside the United States, see Becker & Haag Berlin, *Asbestos; Its Sources, Extraction, Preparation, Manufacture and Uses in Industry and Engineering* (Berlin: Becker & Haag Berlin, 1928), 54.

49. Bruce Yandle, "A Social Regulation Controversy: The Cotton Dust Standard," *Social Science Quarterly* 63, no. 1 (1982): 58–69. Two contrasting views of byssinosis in the cotton textile industry appear in Frank P. Bennett, "It All Depends," *America's Textile Reporter*, July 10, 1969, editorial; and R. C. Davis. "Diagnosis, Brown Lung: Prognosis, Misery," *Textile World* 122 (October 1972): 38–60. See also Rachel Maines, "Brown Lung in Textile Workers," *Off Our Backs* 5, no. 1 (1975): 22. On black lung, now called coal workers' pneumoconiosis (CWP), see Daniel M. Fox and Judith F. Stone, "Black Lung: Miner's Militancy and Medical Uncertainty, 1968–1972," *Bulletin of the History of Medicine* 54, no. 1 (1980): 43–63; and Alan Derickson, *Black Lung: Anatomy of a Public Health Disaster* (Ithaca, N.Y.: Cornell University Press, 1998). On parallels between the coal and asbestos cases, see U.S. Congress, House Committee on Education and Labor, Subcommittee on Compensation Health and Safety, *Asbestos-Related Occupational Diseases: Hearings before the Subcommittee on Compensation, Health, and Safety of the Committee on Education and Labor, House of Representatives, Ninety-Fifth Congress, Second Session* (Washington, D.C.: U.S. General Printing Office, 1979), 125–28.

50. A. W. Rukeyser, "Asbestos Industry Strives to Meet Increasing Demand," *Engineering and Mining Journal* 151 (May 1950): 95, 99; "Plant Designed for Employees' Welfare: Controlling Asbestos Dust at Asten-Hill Manufacturing Company," *Architectural Record* 110 (November 1951): 121–23; and J. Goldfield, "Air Handling and Dust Control in Johns-Manville's New Asbestos Mill," *Mining Engineering* 7 (November 1955): 1029–32.

51. *Minerals Yearbook* (Washington, D.C.: U.S. General Printing Office, 1945), 1457; (1946), 143; (1947), 143; (1948), 144; (1949), 139; (1950), 139; (1951), 167; and (1952), 1:166.

52. Oliver Bowles and F. M. Barsigian, "Asbestos," in *Minerals Yearbook*, ed. Allan F. Matthews (Washington, D.C.: U.S. General Printing Office, 1950), 139.

53. "Have to Have Not Pattern Shown in Strategic Materials Surveys; National Security Board Studies," *Chemical and Engineering News*, August 25, 1952, p. 3536.

54. Robert B. Keating, *Future U.S. Imports of Industrial Raw Materials and Their Ocean Transportation* (Washington, D.C.: National Academy of Sciences, NRC, Committee on Undersea Warfare, 1962), 25.

55. Timothy C. May, "Asbestos," in *Minerals Yearbook* (1961), 1:283; (1964), 1:223–24; and (1967), 197.

56. Horst Mendershausen, "Terms of Trade between the Soviet Union and Smaller Communist Countries, 1955–1957," *Review of Economics and Statistics* 41, no. 2, part 1 (1959): 110; and May, "Asbestos" (1965), 207.

57. Robert S. Moulton, "Interior Finish for Life Safety from Fire," *NFPA Quarterly* 55, no. 1 (1961): 43–50.

58. Underwriters' Laboratories [hereafter UL], *List of Inspected Mechanical Appliances* (Chicago: UL, 1924); UL, *Fire Protection Equipment List, January, 1952* (Chicago: UL, 1952); UL and American Standards Association, *Standard Specifications for Fire Tests of Building Construction and Materials*, 4th ed. (Chicago: UL, 1947); and "Are Your Doors Fire-Safe?" *Safety Maintenance* 132 (November 1966): 33–34.

59. "Asbestos-Cement Products: What They Are and How to Use Them," *American Builder* 72 (October 1950): 83–90; "Asbestos-Cement Panels for Exterior Walls of Stone," *House and Home* 8 (September 1955): 174, 210; Cyrus C. Fishburn, *Physical Properties of Some Samples of Asbestos-Cement Siding*, Building Materials and Structures, report 122 (Washington, D.C.: U.S. General Printing Office, 1951); and Nolan D. Mitchell, *Fire Tests of Wood-Framed Walls and Partitions with Asbestos-Cement Facings*, Building Materials and Structures, report 123 (Washington, D.C.: U.S. General Printing Office, 1951).

60. Asbestos-Cement Products Association, *Advanced Designing with Asbestos Siding: Six Original House Designs by Leading Architects, Utilizing the Color, Texture and Beauty of Asbestos-Cement Siding* (New York: Asbestos-Cement Products Association, 1954), n.p.; Paul W. Icke, "Asbestos," in *Minerals Yearbook* (Washington, D.C.: U.S. General Printing Office, 1968), 180; and Charles L. Readling, "Asbestos," in *Minerals Yearbook* (Washington, D.C.: U.S. General Printing Office, 1969), 188.

61. *Standard Specification for Asbestos-Cement Fiberboard Insulating Panels*, ASTM C551-67; *Standard Definitions of Terms Relating to Asbestos-Cement and Related Products*, ASTM C460-67; *Standard Method of Test for Organic Fiber Content of Asbestos-Cement Products*, ASTM C458-62; *Standard Specifications for Joints for Circular Concrete Sewer and Culvert Pipe, Using Flexible, Watertight Rubber Gaskets*, ASTM C443-67; *Standard Specifications for Asbestos-Cement Nonpressure Sewer Pipe*, ASTM C428-67; *Standard Specifications for Asbestos-Cement Pressure Pipe*, ASTM C296-67; *Standard Specifications for Asbestos-Cement Siding*, ASTM C223-66; *Standard Specifications for Asbestos-Cement Roofing Shingles*, ASTM C222-66; *Standard Specifications and Methods of Test for Corrugated Asbestos-Cement Sheets*, ASTM C221-61; *Standard Specifications for Flat Asbestos-Cement Sheets*, ASTM C220-67; *Standard Methods of Sampling and Testing Asbestos-Cement Flat Sheets, Roofing and Siding Shingles, and Clapboards*, ASTM C459-63; *Standard Methods of Testing Asbestos-Cement Pipe*, ASTM C500-67; and *Standard Specifications for Asbestos-Cement Perforated Underdrain Pipe*, ASTM C508-67.

62. "Roof Coating Materials," *Consumers Research Bulletin* 30 (September 1952): 28–30; and "Asbestos Pouch for Precious Papers," *Consumer Bulletin* 51 (June 1968): 4.

63. W. Francis and J. Grayson, "Some Fundamental Properties of Chrysotile Asbestos in Relation to Electrical Insulation," *Journal of the Society of Chemical Industry* 60 (June 1941): 160–66; W. Lethersich, "Asbestos Covered Cables: Insulation Resistance and Electric Strength," *Electrician*, August 15, 1941, pp. 87–88; NFPA, *National Electrical Code, 1947* (Boston: NFPA, 1947), 84–87; "Properties of Asbestos, Cotton, Duck, Rayon and Silk Insulation," *Electrical World*, August 14, 1950, p. 105; P. O. Nicodemus, "Significance of Iron in Asbestos Materials Used for Electrical Insulating Purposes," *A.S.T.M. Bulletin* 237 (April 1959): 62–67; and "Treated Asbestos Arc-Proofs Manhole Cables," *Electrical World*, July 23, 1956, p. 113.

64. Robert J. Fabian, "Thermal Insulation Materials," *Materials in Design Engineering* 47 (March 1958): 122–24; and "Lagging Cuts Work, Costs and Danger," *Chemical Engineering*, October 9, 1967, p. 132.

65. "Safety Clothing Standards Improved to Protect Workers," *SA* 173 (July 1945): 51; "Safe Suits for Hot Spots: Aluminum-Coated Asbestos Suits," *Business Week*, June 30, 1956, p. 136; "Aluminum Foil on Asbestos: Worn as a Suit, Men Enter Roaring Fires, Etc.," *Iron Age*, December 6, 1956, p. 166; "Light Fabrics Fight Heavy Heat," *Safety Maintenance* 113 (February 1957): 20–21; "Asbestos Gets Tough: New Safety Garments for Molten Metal Handlers," *Safety Maintenance* 128 (November 1964): 25–26; and Bal Dixit, "Performance of Protective Clothing: Development and Testing of Asbestos Substitutes," in *Performance of Protective Clothing: A Symposium Sponsored by ASTM Committee F-23 on Protective Clothing, Raleigh, N.C., 16–20 July 1984*, ed. Roger L. Barker (Philadelphia: ASTM, 1984), 446–60.

66. R. H. Heilman and R. A. MacArthur, "Metal and Asbestos Ducts in Air Conditioning Systems," *Refrigerating Engineering* 37 (February 1939): 105–11; and D. W. French and T. R. Gillen, "Ducts of Asbestos-Cement with Unobstructed Joints and Smooth Surfaces Have 30 Per Cent Lower Pressure Drop," *ASHRAE Journal* 2 (October 1960): 66–67, 110.

67. Jerome Campbell, "Pipe with a Future," *SA* 176 (January 1947): 27–29; Daniel F. Altemus, "Lower Cost Sewer Systems with Asbestos-Cement Pipe," *American City* 72 (December 1957): 106–7; "Realistic Sewer Pipe Design for Asbestos Cement Pipe," *American City* 74 (February 1959): 157, 159; and Keasbey & Mattison Company, *Instructions for Laying "Century" Asbestos-Cement Pipe* (Ambler, Pa.: Keasbey & Mattison Company, 1943).

68. W. Lerch, "Chemical Resistance of Asbestos-Cement Pipe," *Materials Research and Standards* 2 (September 1962): 745–46; P. J. Brennan and N. L. Nemerow, "Studies of Asbestos-Cement Pipe and Plate: Selected Chemical and Mechanical Tests," *Materials Research and Standards* 3 (March 1963): 217–23; G. Klein, "War-Time Water Line Proves Peace-Time Bonus," *Public Works* 96 (February 1965): 93; K. B. Edmunds, "Cost Estimating Building Service Systems; Piping Charts, Asbestos-Cement," *Air Conditioning, Heating and Ventilation* 64 (September 1967): 59–60; and "Underground Asbestos-Cement Piping Carries Chilled Water to Saigon Air Base Complex," *Heating-Piping* 40 (February 1968): 192–93.

69. Ray A. Nixon and Dan Johnson, "Asbestos Asphalt in Atlanta," *American City* 76 (March 1961): 93–94; E. Johannemann, "It Costs Less, Accomplishes More," *American City* 79 (April 1964): 15; and A. M. Johnson, "Why We Add Asbestos," *American City* 81 (March 1966): 98–99.

70. "More Asphalt Streets Use Asbestos," *American City* 78 (May 1963): 9; J. A. Bishop, *Evaluation of Asbestos Asphalt Paving Mixes* (Port Hueneme, Calif.: Naval Civil Engineering Laboratory, 1964); "Asbestos Asphalt Is Useful for Renewal of Decks on Barges and Work Boats," *Marine Engineering/Log* 71 (July 1966): 49; "More Repaving for the Same Cost with Asbestos-Asphalt Mix," *Public Works* 98 (May 1967): 93–94; E. R. Kasper, "Runway Overlay Applied Overnight at Toledo Airport," *Civil Engineering* 39 (August 1969): 58–60; "Bridge Deck Overlay Gains Stability with Asbestos Fibers," *American City* 84 (February 1969): 87–89; and Richard C. Olton, "Asbestos-Asphalt Paving Gives Streets Longer Life," *American City* 88 (September 1973): 105.

71. U.S. Waterways Experiment Station, *Effects of Asbestos Fibers in Asphaltic Concrete Paving Mixtures* (Vicksburg, Miss.: Waterways Experiment Station, 1959); and B. P. Martinez and J. W. Bamberger, "Polyester Fibers Replace Asbestos in Asphalt Curb Mix," *Public Works* 110 (January 1979): 73–74.

72. "Lightweight Roofing Has Aluminum Foil Vaporbarrier, Asbestos Fibers for Insulation," *Architectural Forum* 92 (June 1950): 182, 184; "New Uses Found for Asbestos-Cement in Research; Oklahoma City Research House," *Progressive Architecture* 41

(February 1960): 73; "Sculptured Asbestos Cement," *Progressive Architecture* 52 (September 1971): 146–51; and "Asbestos Buildings," *Architectural Design* 42 (November 1972): 717–18. On uses in low-cost housing for developing countries, see U. S. International Cooperation Administration, Technical Aids Branch, *Plant Requirements for Manufacture of Asbestos-Cement Pipe* (Washington, D.C.: George Andrews Engineering Associates, 1960); and the United Nations Technical Assistance Board's sponsorship of Alvaro Ortega's design, "Pleated Asbestos Roof for Low-Cost Housing," *Architectural Record* 129 (February 1961): 175.

73. National Board of Fire Underwriters [hereafter NBFU], *Property Insurance Fact Book* (New York: NBFU, 1959), 10, 13, 23.

74. Everett Uberto Crosby and Henry Anthony Fiske, *NFPA Handbook of Fire Protection*, 10th ed. (Boston: NFPA, 1948).

75. NFPA, *National Fire Codes* (Boston: NPFA, 1951), 3:274–93, 368–69, 389, 607. Asbestos was standard insulation behind chimney baseboards in Britain as well; see Anthony Medlycott, *Applied Building Construction*, 2d ed. (London: Chapman and Hall, 1957), 1:48, 55.

76. Robert S. Moulton, *NFPA Handbook of Fire Protection*, 11th ed. (Boston: NFPA, 1954): 131, 193, 381, 392, 534–37, 567, 598–601, 615, 622, 642, 646–52, 805, 809, 1241.

77. NFPA, *National Electrical Code, 1971* (Boston: NFPA, 1971), chap. 70, sec. 100, para. 113. The 1947 code had similar recommendations.

78. NBFU, *National Building Code* (New York: NBFU, 1955), 33–37, 163, 227, 232–55, and appendix, pp. 1–43; and NBFU, *Fire Prevention Code* (New York: NBFU, 1960), 19, 63, 162. See also New York City, *Building Code: Local Law No. 76 of the City of New York. Effective December 6, 1968; Amended to August 22, 1969* (New York: City of New York, 1969).

79. NBFU, *Standard Schedule for Grading Cities and Towns of the United States, with Reference to their Fire Defenses and Physical Conditions* (New York: NBFU, 1956), 89, 100; and NBFU, *Fire Hazards and Safeguards for Metalworking Industries*, Technical Survey, no. 2. (New York: NBFU, 1954), 11, 19, 21–22, 46.

80. Building Officials' Conference of America, *Building Code, Compiled by State of Connecticut* (Hartford: Connecticut Public Works Department, Housing Division, 1951), 31, 38–39, 59, 64, 74.

81. International Conference of Building Officials, *Uniform Building Code* (Pasadena, Calif.: International Conference of Building Officials, 1958, 1988).

82. Commonwealth of Massachusetts, *Building Code, Commonwealth of Massachusetts* (Boston: Department of Public Safety, 1960), 17, 50–65, 82–84; New Jersey Bureau of Housing and Bureau of Planning and Commerce, *Manual for the Standard Building Code of New Jersey* (Trenton: Department of Conservation and Economic Development, Division of Resource Development, Bureau of Housing, 1965), 36, 38–39, 63, 68, 109.

83. William H. Correale, *A Building Code Primer* (New York: McGraw-Hill, 1979), 113–28; and Richard D. Peacock, *A Review of Fire Incidents, Model Building Codes, and Standards Related to Wood-Burning Appliances: Final Report, May 1979*, NBSIR 79-1731 (Washington, D.C.: U.S. Department of Commerce, National Bureau of Standards, 1979).

84. Joseph J. Loftus, *Evaluation of Wall Protection Systems for Wood Heating Applicances*, NBSIR 82-2506 (Washington, D.C.: U.S. Department of Commerce, National Bureau of Standards, 1982), 4

85. See, for example, *Life Hazard Test of Gypsum Wallboard from Gypsum Association, Chicago, Illinois* (Boston: Factory Mutual Laboratories, 1951), 1–2 and appendix, sheet 1.

86. Askarel insulating liquids are toxic and are no longer manufactured in the United States. The *Exxon Encyclopedia of Lubricants* defines them as "a group of synthetic, fire-resistant, chlorinated aromatic hydrocarbons;" see *http://www.prod.exxon.com/exxon_productdata/lube_encyclopedia/index.htm*, accessed on January 10, 2004.

87. Associated Factory Mutual Fire Insurance Companies, Engineering Division, *Handbook of Industrial Loss Prevention; Recommended Practices for the Protection of Property and Processes against Damage by Fire, Explosion, Lightning, Wind, Earthquake*, McGraw-Hill Insurance Series (New York: McGraw-Hill, 1959), chap. 1, pp. 6, 11; chap. 2, pp. 5, 9; chap. 5, p. 1; chap. 7, pp. 11, 13; chap. 9, p. 3; chap. 10, pp. 3–4, 7, 8, 10, 14; chap. 13, p. 2; chap. 14, pp. 9, 11; chap. 15, pp. 6–7; chap. 25, p. 6; chap. 26, pp. 1, 4–5; chap. 28, p. 10; chap. 29, pp. 4, 10–11, 13; chap. 34, pp. 5, 12; chap. 35, p. 3; chap. 38, pp. 4, 7–8; chap. 39, pp. 3, 6; chap. 40, pp. 3, 16, 22–23; chap. 42, p. 4; chap. 44, pp. 2, 7; chap. 46, p. 3; chap. 52, p. 2; chap. 54, p. 4, chap. 55, p. 3; chap. 57, pp. 4, 6, 12; chap. 60, pp. 7, 15; chap. 66, pp. 2, 4; chap. 68, p. 1; chap. 69, p. 18; chap. 70, p. 2; chap. 71, p. 2; chap. 74, p. 1. In the 1967 edition, see chap. 5, pp. 1, 4, 8–10, 12; chap. 6, pp. 3, 5, 9; chap. 7, pp. 5, 15, 17; chap. 12, p. 16; chap. 15, pp. 3, 12; chap. 27, pp. 1, 8; chap. 28, pp. 3–5; chap. 34, pp. 8, 10, 13, 15; chap. 38, p. 12; chap. 39, p. 2; chap. 40, p. 4; chap. 43, p. 9; chap. 44, pp. 6–7; chap. 45, pp. 4, 7, 18, 25; chap. 47, p. 2; chap. 49, pp. 7–8; chap. 52, p. 8; chap. 56, p. 2; chap. 58, p. 2; chap. 61, p. 3; chap. 63, pp. 1, 3–4, 8; chap. 64, p. 1; chap. 66, p. 13; chap. 69, p. 17; chap. 70, p. 8; chap. 74, p. 1; chap. 75, pp. 17–18; chap. 76, pp. 1, 3–4; chap. 77, p. 4.

88. NRC, *A Program for Building Research in the United States; A Report for the National Bureau of Standards. Special Advisory Committee on the Desirable Role of the National Bureau of Standards in Building Research*, NRC publication no. 994 (Washington, D.C.: National Academy of Sciences, NRC, 1962); D. Gross, *Flame Spread Properties of Interior Finishes*, report no. 4757 (Washington, D.C.: U.S. Department of Commerce, National Bureau of Standards, n.d.); Jin B. Fang, *Fire Buildup in a Room and the Role of Interior Finish Materials*, NBS Technical Note no. 879 (Washington, D.C.: U.S. Department of Commerce, National Bureau of Standards, 1975); and William J. Parker, *An Investigation of the Fire Environment in the ASTM E-84 Tunnel Test* (Washington, D.C.: U.S. Department of Commerce, National Bureau of Standards, Institute for Applied Technology, Center for Building Technology, 1977).

89. Walter Charles Voss, *Fireproof Construction* (New York: Van Nostrand, 1948), 267; Louis Przetak, *Standard Details for Fire-Resistive Building Construction* (New York: McGraw-Hill, 1977): 11, 147, 156–57, 187, 203, 247, 264, 268, 286, 291; James J. Williamson, *General Fire Hazards and Fire Prevention*, 5th ed., Chartered Insurance Institute Handbook no. 8 (London: Pitman, 1965), 31–33; M. David Egan, *Concepts in Building Firesafety* (New York: Wiley, 1978), 96, 104; Bertram Le Roy Wood, *Fire Protection through Modern Building Codes*, 4th ed. (New York: American Iron and Steel Institute, 1971): 63, 75; Eric Walter Marchant, *A Complete Guide to Fire and Buildings* (Lancaster, U.K.: Medical and Technical Publishing, 1972; reprint, Barnes and Noble, 1973), 121, 133; Nihon Kasaigakkai [Japanese Association of Fire Science and Engineering], *Evaluation of Fire Safety in Buildings*, Occasional Report of the Japanese Association of Fire Science and Engineering, no. 3 (Tokyo: Nihon Kasaigakkai, 1979), 143; and Philip S. Schaenman, *International Concepts in Fire Protection: Ideas from Europe That Could Improve U.S. Fire Safety* (Arlington, Va.: TriData, 1982), esp. 1–35.

90. U.K. Department of Scientific and Industrial Research, Fire Offices' Committee, Joint Fire Research Organization, *Flame-Retardant Building Materials* (London: Her Majesty's Stationery Office, 1959), 3–8; and its *Sponsored Fire-Resistance Tests on Structural Ele-*

ments (London: Her Majesty's Stationery Office, 1960): 2, 7, 18, 30, 42–43, 51, 54–55, 67–69. For later works under the U.K. Department of the Environment, see P. L. Hinkley, *The Fire Propagation Test as a Measure of the Fire Hazard of a Ceiling Lining*, ed. H.G.H. Wraight and Ann Wadley (Boreham Wood, U.K.: Fire Research Station, 1968): 1 and fig. 6; C. R. Theobald, *The Effect of Roof Construction and Contents on Fires in Single Storey Buildings*, Fire Research Note no. 941 (Boreham Wood, U.K.: Fire Research Station, 1972), 9–11 and fig. 2; and W. D. Woolley, *Performance of Asbestos Fire Blankets*, ed. S. P. Rogers (Boreham Wood, U.K.: Fire Research Station, 1976).

91. For U.S. civilian defense policy during World War II regarding the role of air-raid shelters in conventional warfare, see U.S. Interdepartmental Advisory Committee on Protection, and Public Buildings Administration, *Air Raid Protection Code for Federal Buildings and Their Contents* (Washington, D.C.: U.S. General Printing Office, 1942); on nuclear civil defense, see Diane Diacon, *Residential Housing and Nuclear Attack* (London: Croom Helm, 1984), 21, 28, 58–59, 78, 131.

92. On the nuclear heat wave, see Samuel Glasstone, *The Effects of Nuclear Weapons*, rev. ed. (Washington, D.C.: U.S. Atomic Energy Commission, 1962), 316–68.

93. George J. B. Fisher, *Incendiary Warfare* (New York: McGraw-Hill, 1946), 88; U.S. Commission on Intergovernmental Relations, *A Staff Report on Civil Defense and Urban Vulnerability, Submitted to the Commission on Intergovernmental Relations* (Washington, D.C.: U.S. General Printing Office, 1955), 9, 21; T. E. Lommasson, *Nuclear War and the Urban Fire Problem* (Albuquerque, N.M.: Dikewood Corporation for the U.S. Office of Civil Defense, 1964), 52; R. H. Renner, S. B. Martin, and R. E. Jones, *Parameters Governing Urban Vulnerability to Fire from Nuclear Bursts (Phase 1)* (San Francisco: U.S. Naval Radiological Defense Laboratory, 1966), 128; and Sargent-Webster-Crenshaw and Folley, *Reducing Urban Vulnerability to Nuclear Attack through Preventive Planning*, prepared for U.S. Department of Defense, Office of Civil Defense, contract no. OCD-OS-63-200 (Syracuse, N.Y.: Sargent-Webster-Crenshaw and Folley, 1964), 2–5.

94. U. S. Office of Civil Defense, *Family Shelter Designs* (Washington, D.C.: 1962), 7–10; see also "Civil Defense: The Sheltered Life, *Time,* October 20, 1961, pp. 21–26.

CHAPTER 6 THE ASBESTOS TORT CONFLAGRATION

1. Paul Brodeur, *Outrageous Misconduct: The Asbestos Industry on Trial* (New York: Pantheon, 1985), 137.

2. Icke, "Asbestos" (1968), 188.

3. For overviews of this literature, see David Ozonoff, "Failed Warnings: Asbestos-Related Disease and Industrial Medicine," in *The Health and Safety of Workers,* ed. R. Bayers (New York: Oxford University Press, 1988); Philip E. Enterline, "Changing Attitudes and Opinions Regarding Asbestos and Cancer 1934–1965," *American Journal of Industrial Medicine* 20, no. 5 (1991): 685–700; and, of course, Castleman and Berger, *Asbestos,* 1–135.

4. S. B. McPheeters, for example, asserted in 1936 that the result of his survey of asbestos workers "warrants the opinion that with the reduction of dust concentration now feasible by available dust eliminating devices, the incidence of significant asbestosis would in time approach the vanishing point." See "A Survey of a Group of Employees Exposed to Asbestos Dust," *Journal of Industrial Hygiene and Toxicology* 18 (1936): 239.

5. Groff Conklin, "Cancer and Environment," *SA* 180, no. 1 (1949): 11–15.

6. NRC, *Asbestos*, 6.

7. Robert A. Clifton, "Asbestos," in *Minerals Yearbook* (Washington, D.C.: U.S. General Printing Office, 1978–79), 71–73; (1983), 1:113–14; and Robert La Virta, "Asbestos," in *Minerals Yearbook* (Washington, D.C.: U.S. General Printing Office, 1988), 133; (1990), 172. The Court upheld the Environmental Protection Agency's right to ban new uses of asbestos but not those that historically contained the mineral. U.S. Environmental Protection Agency [hereafter EPA], "Asbestos; Manufacture, Importation, Processing and Distribution Prohibitions; Effect of Court Decision," 40 CFR, part 763, *Federal Register* 57, no. 64 (1992): 11364–65. On brakes, see American Society of Mechanical Engineers Expert Panel on Alternatives to Asbestos in Brakes and EPA, *Analysis of the Feasibility of Replacing Asbestos in Automobile and Truck Brakes* (New York: American Society of Mechanical Engineers, 1988).

8. *Mealey's Litigation Report: Asbestos* (King of Prussia, Pa.: Mealey, 1993– present); and *Mealey's Asbestos Bankruptcy Report* (King of Prussia, Pa.: Mealey, 2001–present). A third publication by Mealey, *International Asbestos Liability*, first appeared in 2003.

9. Walter E. Dellinger, "Asbestos Claims Trust: Testimony of Prof. Walter E. Dellinger" (2002), *http://www.LitigationDataSource.com,* accessed on November 15, 2003.

10. D. E. Lilienfeld et al., "Projection of Asbestos-Related Diseases in the United States, 1985–2009. I. Cancer," *British Journal of Industrial Medicine* 45, no. 5 (1988): 283–91; and "Report Says Asbestos Deaths Growing into Epidemic," *Medical Letter on the CDC and FDA,* April 18, 2004, p. 30.

11. American Law Institute, *Restatement of the Law Second, Torts* (Philadelphia: American Law Institute, 1965); and Adam Raphael, *Ultimate Risk: The Inside Story of the Lloyd's Catastrophe* (New York: Four Walls Eight Windows, 1995).

12. Castleman and Berger, *Asbestos*, 128, 323.

13. "Asbestos Health Question Perplexes Experts," *Chemical and Engineering News,* December 10, 1973, pp. 18–19; and Ronald E. Gots, *Toxic Risks: Science, Regulation and Perception* (Boca Raton, Fla.: Lewis, 1993), 156, 210–12.

14. Many parallels with the asbestos situation will be found in Christian Warren's *Brush with Death: A Social History of Lead Poisoning* (Baltimore: Johns Hopkins University Press, 2000), but lead toxic torts have not generated an international crisis on the scale of asbestos torts. On the toxic effects of lead on children, see EPA, "Lead in Paint, Dust and Soil," *http://www.epa.gov/lead/index.html,* accessed on May 3, 2004.

15. The Rand Institute for Civil Justice has been tracking the costs of asbestos litigation since 1983. See, for example, James S. Kakalik, *Costs of Asbestos Litigation* (Santa Monica, Calif.: Rand Institute for Civil Justice, 1983); James S. Kakalik et al., *Variation in Asbestos Litigation Compensation and Expenses* (Santa Monica, Calif.: Rand Center for Civil Justice, 1984); and Molly Selvin and Larry Picus, *The Debate over Jury Performance: Observations from a Recent Asbestos Case* (Santa Monica, Calif.: Rand Institute for Civil Justice, 1987).

16. CDC, Center for Infectious Diseases, "Food Related Diseases," *http://www.cdc.gov/ncidod/diseases/food/index.htm,* accessed on May 12, 2004.

17. Brodeur, *Outrageous Misconduct,* 6. Brodeur is a journalist who specializes in malevolent caricatures of American industry; electric companies and manufacturers of microwave appliances and cell phones as well as asbestos have all been em-Brodeured.

18. Gustave Shubert, Foreword, in *Asbestos in the Courts: The Challenge of Mass Toxic Torts,* by Deborah R. Hensler et al. (Santa Monica, Calif.: Rand Center for Civil Justice, 1985), iii–iv.

19. Sheila Jasanoff, *Science at the Bar: Law, Science, and Technology in America* (Cambridge, Mass.: Harvard University Press, 1995), 15, 30, 117–23.

20. Castleman and Berger, *Asbestos; C. Borel v. Fiberboard Paper Products et al.*, 493 Fed. 2d 1076 U.S. Circuit Court of Appeals for the 5th Circuit, September 10, 1973; and Richard A. Epstein, "Implications for Legal Reform," in *Regulation through Litigation*, ed. W. Kip Viscusi (Washington, D.C.: American Enterprise Institute–Brookings Joint Center for Regulatory Studies, 2002), 343–47. It is also illuminating to view Castleman and Berger as a kind of draft script in the theatrical metaphor suggested by Stephen Hilgartner, who examines the presentation of medical and scientific opinion in the context of health controversies within federal agencies in *Science on Stage: Expert Advice As Public Drama* (Stanford, Calif.: Stanford University Press, 2000).

21. Some of them apparently eat very well; see Jonathan D. Glater, "A Houston Holiday: Barbecue, Al Green and 5,000 Guests," *New York Times*, December 13, 2003, pp. B1, B3.

22. See, for example, D. E. Lilienfeld, "The Silence: The Asbestos Industry and Early Occupational Cancer," *American Journal of Public Health* 81 (June 1991): 791–800.

23. Francis H. May, "Prepared Statement of Francis H. May, Executive Vice President, Johns-Manville Corp.," quoted in Castleman and Berger, *Asbestos*, 897.

24. The first year in which asbestosis is so listed is 1930, although there is an entry for "Asbestosis—See Lung—Dust diseases" in the 1926–27 volume. See H. W. Wilson Company, *Industrial Arts Index* (New York: H. W. Wilson Company, 1926–27), 132; (1930), 102. After 1958 this title splits into the *Applied Science and Technology Index* and the *Business Periodicals Index*.

25. As of 1958, at least five hundred American libraries had subscriptions to the *Industrial Arts Index*; sixty-two major repositories were listed in the Library of Congress's *National Union Catalog Pre-1956 Imprints*, 18:642. It would not even have been necessary to read the articles because the titles of the relevant studies indicate that asbestos was a known inhalation hazard and suspected carcinogen.

26. Geoffrey Tweedale, *Magic Mineral to Killer Dust: Turner & Newall and the Asbestos Hazard* (New York: Oxford University Press, 2000), 249.

27. "Can You Afford to Be Hurt?" *Asbestos Worker* 12 (October 1944): 16; but see, for example, cover illustrations, 15, no. 2 (January 1958); and no. 10 (February 1960). Also see Bill Sells's comments on workers' fears about their jobs in "What Asbestos Taught Me about Managing Risk," in *Harvard Business Review*, reprint no. 94209 (Boston: Harvard Business School, March–April 1994), 7.

28. Castleman and Tweedale both note this dislike of face masks on the part of workers; see Castleman and Berger, *Asbestos*, 5, 232, 332; and Tweedale, *Magic Mineral*, 128, 202. W. Fleischer et al. saw no asbestos workers wearing the required respirators in the Navy shipyards they inspected in 1945; see their "A Health Survey of Pipe Covering Operations in Constructing Naval Vessels," *Journal of Industrial Hygiene and Toxicity* 28 (1946): 9–16. See also Philip Drinker, "Health and Safety in Contract Shipyards during the War," *Occupational Medicine* 3, no. 4 (1947): 335–43; and Jacqueline Karnell Corn and Jennifer Starr, "Historical Perspective on Asbestos: Policies and Protective Measures in World War II Shipbuilding," *American Journal of Industrial Medicine* 11 (1987): 359–73. On the inutility of warnings, see Mary Douglas and Aaron B. Wildavsky, *Risk and Culture: An Essay on the Selection of Technical and Environmental Dangers* (Berkeley: University of California Press, 1982), 75–77.

29. NRC, *Asbestos*, 26.

30. On the history of product warning labels, see James M. Miller and Mark R. Lehto, *Instructions and Warnings: The Annotated Bibliography* (Ann Arbor, Mich.: Fuller, 1994), vii. Ammonium nitrate is not ordinarily combustible. See the remarks of E. H. Whittemore, then state fire marshal of Massachusetts, in "Report of Committee on Flammable Liquids," *Proceedings of the NFPA Fifty-first Annual Meeting*, 92.

31. Conklin, "Cancer and Environment," 14

32. Marty Ahrens, *Wood Shingle or Wood Shake Roof Fires: Statistical Analysis* (Quincy, Mass.: NFPA, 2001), 3. The author notes that in the late 1990s this annual average was decreasing.

33. W. C. Hueper, "Experimental Studies in Metal Cancerigenesis. VI. Tissue Reactions in Rats and Rabbits after Parenteral Introduction to Suspensions of Arsenic, Beryllium, or Asbestos in Lanolin," *Journal of the National Cancer Institute* 15 (1955): 113–29. On rabbits, see E. J. King and J. W. Clegg, "Effect of Asbestos and of Asbestos and Aluminum on Lungs of Rabbits," *Thorax* 1 (1946): 188–97; five years later King published on rats in the same journal with J. M. McLean-Smith and D. P. Wooton, 6 (1951): 127–36. See also Arthur J. Vorwald, T. M. Durkan, and P. C. Pratt, "Experimental Studies of Asbestosis," *Archives of Industrial Hygiene and Occupational Medicine* 3 (1951): 1–43. On mice and rats, see K. M. Lynch et al., "Pulmonary Tumors in Mice Exposed to Asbestos Dust," *AMA Archives of Industrial Health* 15 (March 1957): 207–14; A. G. Bobkov, "Zlokachestvennaia Mezotelioma Pridatkov Iaichka, Vyzvannaia u Beloi Krysy 9, 10–Dimetil–1, 2–Benzantratsenom," *Voprosy Onkologii* 11, no. 8 (1965): 100–102; J. C. Wagner and J. W. Skidmore, "Asbestos Dust Deposition and Retention in Rats," *Annals of the N.Y. Academy of Sciences* 132, no. 1 (1965): 77–86; G. E. Westlake, H. J. Spjut, and M. N. Smith, "Penetration of Colonic Mucosa by Asbestos Particles: An Electron Microscopic Study in Rats Fed Asbestos Dust," *Laboratory Investigations* 14, no. 11: 2029–33; P. Gross et al., "Experimental Asbestosis: The Development of Lung Cancer in Rats with Pulmonary Deposits of Chrysotile Asbestos Dust," *Archives of Environmental Health* 15, no. 3 (1967): 343–55; J. M. Jagatic et al., "Tissue Response to Intraperitoneal Asbestos with Preliminary Report of Acute Toxicity of Heart-Treated Asbestos in Mice," *Environmental Research* 1, no. 3 (1967): 217–30; F. J. Roe et al., "The Pathological Effects of Subcutaneous Injections of Asbestos Fibres in Mice: Migration of Fibres to Submesothelial Tissues and Induction of Mesotheliomata," *International Journal of Cancer* 2, no. 6 (1967): 628–38; and R. L. Carter, "Pathology of Ovarian Neoplasms in Rats and Mice," *European Journal of Cancer* 3 (1968): 537–43.

34. For example, I was denied access to a marketing report performed for the National Home Sewing Association in 1977 on the grounds that it was proprietary. The contents, it was whispered, included the prediction of a flat (that is, no-growth) market for home sewing products through the 1980s. I have since been denied access to unpublished proprietary reports on the subjects of sex toy sales, marketing of hotel brands, and the feasibility of credit-card-guaranteed restaurant reservations.

35. Arthur J. Vorwald and John W. Karr, "Pneumoconiosis and Pulmonary Carcinoma," *American Journal of Pathology* 14 (1938): 49–57.

36. Arthur J. Vorwald, T. M. Durkan, and P. C. Pratt, "Experimental Studies of Asbestosis," *Archives of Industrial Hygiene and Occupational Medicine* 3 (1951): 41; and Philip Ellman, "Pneumoconiosis: Part III—Pulmonary Asbestosis," *British Journal of Radiology* 7 (1934): 287–88.

37. John Anderson and Francis A. Campagna, "Asbestos and Carcinoma of the Lung: Case Report and Review of the Literature," *Archives of Environmental Health* 1 (July 1960): 39–44.
38. The Sumner Simpson correspondence is reproduced in full in Castleman and Berger, *Asbestos*, 879–82; see also Brodeur, *Outrageous Misconduct*, 116–17.
39. Some fire-safety professionals would argue that we still have not achieved an acceptable level because our fire losses are still larger than those of most other industrial nations. See Karter, *Fire Loss*.
40. *Action Program: The President's Conference on Fire Prevention, the President's Conference on Fire Prevention; Variation: House Document,* U.S. Congress, House, report no. 17920, 4 (Washington, D.C.: U.S. General Printing Office, 1947), 1; and Castleman and Berger, *Asbestos*, 107.
41. "Asbestos Is Unique and Indispensable," *Engineer,* April 24, 1969, p. 40; L. McLain, "Seeking Alternatives to Hazardous Asbestos," *Engineer,* May 20, 1976, pp. 30–31, 34; L. McGinty, "Risk Equations: A Ban on Asbestos?" *New Scientist,* July 14, 1977, pp. 96–97; and N. Gupta, "Searching for the Asbestos Substitute," *Engineer,* November 22, 1979, p. 41.
42. Robert A. Clifton, "Asbestos," in *Minerals Yearbook* (Washington, D.C.: U.S. General Printing Office, 1973), 171; and Joan C. Todd, "Asbestos, 1974: Demand May Exceed Supply in 1975," *Engineering and Mining* 176 (March 1975): 144–45.
43. Castleman and Berger, *Asbestos*, 103.
44. Benzaquin, *Fire in Boston's Cocoanut Grove,* 221.
45. See, for example, Belluck and Fox, LLP, "Why Was Asbestos Used in Building and Insulation Products?" *http://www.belluckfox.com/asbestos08.html,* accessed on January 29, 2003.
46. Peter W. Bartrip, "Irving John Selikoff and the Strange Case of the Missing Medical Degrees," *Journal of the History of Medicine and Allied Sciences* 58, no. 1 (2003): 3–33; and Joseph N. Gitlin et al., "Comparison of 'B' Readers' Interpretations of Chest Radiographs for Asbestos Related Changes," *Academic Radiology* 11 (August 2004): 843–56.
47. Sells, "What Asbestos Taught Me," 4.
48. Peter Bartrip notes this problem in *The Way from Dusty Death: Turner and Newall and the Regulation of Occupational Health in the British Asbestos Industry, 1890s–1970* (London: Athlone, 2001), 1.
49. Castleman and Berger, *Asbestos*, 363, 793.
50. I have already cited much of this documentation in chapter 4; but see U.S. War Production Board, *Conservation Order No. M-79 as Amended June 18, 1942* (Washington, D.C.: U.S. Government Printing Office, 1942), which is entirely unambiguous about the requirement that asbestos be used in the fulfillment of defense orders, particularly those of the Navy.
51. "Special Report: Airbag Safety," *Consumers' Digest* 36, no. 2 (1997): 21; and Sharon Collins, "Do Flame-Retardants Save Lives or Endanger Children?" *CNN.com* (2003), *http://www.cnn.com/2003/TECH/science/05/05/hln.hot.earth.pbde/index.html,* accessed on May 5, 2003. On the inevitability of risk, see Philip Alcabesm, "Epidemiologists Need to Shatter the Myth of a Risk-Free Life," *Chronicle of Higher Education,* May 23, 2003, pp. B11–12.
52. "More Dangers from Asbestos and Its Substitutes," *New Scientist,* April 8, 1982, p. 86.

53. Therese McAllister et al., *World Trade Center Building Performance Study: Data Collection, Preliminary Observations, and Recommendations* (Washington, D.C.: Federal Emergency Management Agency, Federal Insurance and Mitigation Administration, 2002), 2–11.

54. Thomas W. Eagar and Christopher Musso, "Why Did the World Trade Center Collapse? Science, Engineering and Speculation," *Journal of Materials* 53, no. 12 (2001): 8–11.

55. The foregoing account relies on Witold Rybczynski, "What We Learned About Tall Buildings from the World Trade Center Collapse," *Discover* 23, no. 10 (2002): 68–76; Zdenk P. Bazant and Yong Zhou, "Why Did the World Trade Center Collapse?—Simple Analysis" (2001), *http://www.tam.uiuc.edu/new/200109wtc/*, accessed on January 12, 2004; James Glanz and Andrew C. Revkin, "Haunting Question: Did the Ban on Asbestos Lead to Loss of Life?" *New York Times*, September 18, 2001, p. F2; CDC, National Institute for Occupational Safety and Health, *Summary Report to the New York City Department of Health: NIOSH Air Sample Results for the World Trade Center Disaster Response* (Atlanta: CDC, National Institute for Occupational Safety and Health, 2002); Roger N. Clark et al., *Images of the World Trade Center Site Show Weak Patterns of Minerals with Spectral Absorptions Indicating the Possible Presence of Asbestiform Minerals* (Washington, D.C.: U.S. Geological Survey, 2003); EPA, *Health Effects of the World Trade Center Collapse* (Washington, D.C.: EPA, 2003); U.S. National Health and Environmental Effects Research Laboratory, *Toxicological Effects of Fine Particulate Matter Derived from the Destruction of the World Trade Center* (Research Triangle Park, N.C.: EPA, 2002), 44; Vincent Dunn, "Why the World Trade Center Buildings Collapsed: A Fire Chief's Assessment," *http://www.vincentdunn.com/wtc.html*, accessed on January 18, 2003; Laurie Kazan-Allen, "Asbestos Use in the Construction of the World Trade Center" (2001), *http://www.btinternet.com/~ibas/lka_world_trade_center.htm*, accessed on January 18, 2003; Steven Milloy, "Asbestos Could Have Saved WTC Lives," *Fox News Channel* (2001), *http://www.foxnews.com/printer_friendly_story/0,3566,34342,00.html*, accessed on January 16, 2003; "World Trade Centre [*sic*]—New York—Some Engineering Aspects," *University of Sydney, Department of Civil Engineering* (2001), *http://www.civil.su.oz.au/wtc_enr.htm*, accessed on January 18, 2003; Steven Ashley, "Collapse of the World Trade Center Towers," *SA* (2001), *http://www.sciam.com/article.cfm?articleID=000B7FEB-A88C-1C75-9B81809EC588EF21&catID=4*, accessed on January 12, 2003; Dave Williams, "The Bombing of the World Trade Center in New York City," *International Criminal Police Review*, nos. 469–71 (1998): 469–71; and Wirtualna Polska, "Briefcase Saved Lives of Polish Army Officers in the Pentagon," *Polish Agency of Information* (2002), *http://www.polishnews.pl/USA/a_briefcase_saved_lives_of_polis.htm*, accessed on March 11, 2003. It is likely that those "asbestos" curtains were actually made of fiberglass, but it is understandably difficult to obtain confirmation of Pentagon construction details.

56. Chauncey Starr, "Social Benefit versus Technological Risk," *Science* 165, no. 19 (1969): 1232–38.

57. Douglas and Wildavsky, *Risk and Culture*, 28.

58. Edmund Russell, *War and Nature: Fighting Humans and Insects with Chemicals from World War I to Silent Spring*, Studies in Environment and History (Cambridge, U.K.: Cambridge University Press, 2001), 113.

59. Robert Proctor, *Cancer Wars: How Politics Shapes What We Know and Don't Know about Cancer* (New York: Basic Books, 1995), 133–73, 249–70.

60. Douglas and Wildavsky, *Risk and Culture,* 10–11, 128–51.

61. Keyes, *Cocoanut Grove*, 212.

62. See, for example, the Conyers bill (HR 676) endorsed by Physicians for a National Health Plan, described at *http://www.pnhp.org/nhibill/nhi_execsumm.html,* accessed on November 17, 2003, which is only one of many examples of proposed plans.

INDEX

ABOUT THE AUTHOR

RACHEL MAINES is the author of *The Technology of Orgasm: "Hysteria," the Vibrator, and Women's Sexual Satisfaction* (1999), which won the Herbert Feis Award of the American Historical Association. She holds a doctorate in applied history from Carnegie-Mellon University (1983) and is employed in the Department of Applied Economics and Management at Cornell University.